Petroleum Refineries

Petroleum Refineries

A Troubleshooting Guide

Dhananjoy Ghosh

JENNY STANFORD
PUBLISHING

Published by

Jenny Stanford Publishing Pte. Ltd.
Level 34, Centennial Tower
3 Temasek Avenue
Singapore 039190

Email: editorial@jennystanford.com
Web: www.jennystanford.com

British Library Cataloguing-in-Publication Data
A catalogue record for this book is available from the British Library.

Petroleum Refineries: A Troubleshooting Guide

ISBN 978-981-4877-81-7 (Hardcover)
ISBN 978-1-003-18040-1 (eBook)

This book is dedicated to the memories of my heavenly parents.

Contents

Acknowledgment

As a young chemical engineer in a petroleum refinery, I came across with possession of a book on troubleshooting authored by the renowned engineer, Norman P. Lieberman. The way he documented the book, it was amply clear that the troubleshooting experiences that we earn in our profession are worth documenting if these are expressed with logics, arguments and illustrations like he did, and this can be an excellent knowledge bank for future engineers and will also help refineries to advance in the steps of profitability and sustainability in operations. Writing this book was a natural inspiration from his work.

My colleagues were well aware of my apt in troubleshooting methodologies and achievement. Hence, they were always vocal in suggesting to me to write such a book that would obviously help the refinery sectors. They also inspired me a lot to write this book.

Last but not the least, continuous inspiration from my wife, Mita, and her support in writing this book has made it possible to complete the mission.

About the Author

D. Ghosh is an industrial professional and visiting faculty in universities. He is a chemical engineer and also did his graduation with honours in chemistry; he holds a PhD in petroleum technology, all from Calcutta University, India. Since the beginning, his focus was on innovation, troubleshooting and safety. He worked with several refineries for safe commissioning of their projects, and in this way, he became an expert of operation in all petroleum technologies. Due to his apt in innovation, he developed technology of a new process, "re-extraction of Aromatics from petroleum residual extract" in 1995 and got an innovation award from NPMP (under the Petroleum Ministry of India) in 2000. In 1996, he developed another technology, "simultaneous deoiling and dewaxing in a single unit to produce quality base oil and micro-crystalline wax at the same time" with the objective of integration and energy savings, which was commercially implemented in 2002. He has authored a book on "Safety in Petroleum Industries" published by CRC Press, Taylor & Francis Group. He has a number of research publications in national and international journals. He served petroleum refineries for about 35 years reaching senior positions and heading refinery operations, maintenance, safety and training. He also served chemical and petrochemical industries in senior positions after a long stint in the refinery sector. He worked as an expert panel member at the Centre for High Technology, India, on behalf of the Petroleum Ministry of India for the audit of benchmarking performance of Indian refineries.

Preface

Due to being an industry professional engaged with multiple organizations in the refinery sector, I got enormous opportunities to get experienced in handling practical problems of all technologies in refineries, their operations, safety and maintenance. Learning the subject with passion, I used to demonstrate the knowledge to the engineers, field operators of different education levels and understanding capabilities engaged in the operation.

This book is a simple compilation of the above narrative with the objective to provide a knowledge bank for the present and future engineers and university students to find them easy to handle the upset, if any, in the process plants safely and without loss of time thus enhancing the availability of plants resulting in profitability and safety. With the knowledge from this book, students and operating engineers should be able to contribute more to the development of process design improvement, enhancing standard operating procedures (SOPs) and enhancing safety features in operational instrumentation and practices.

The chapters on troubleshooting methodologies and key issues in troubleshooting have been discussed with a recap on unit operation principles in operating a process plant.

The book provides an enormous number of case studies—more than 200—in two categories: (1) equipment category, where operations mostly with all types of equipment used in refineries have been covered, and (2) troubleshooting, which covers operations with all prevailing technologies in refineries. Technology coverage has been kept such that refinery configurations may be different across different organizations but operators will find the applicable case studies in this book. Methodologies, reasoning, sequencing and illustrations are provided in a way to make the book user-friendly.

Chapter 1

Introduction

1.1 Definitions

Troubleshooting is a common affair the individuals or organizations exercise for their survival.

In petroleum refineries also, though there are sets of guidelines/SOPs to operate the plants, sometimes unique problems arise which need to be tackled with knowledge, experience and ready wittiness to overcome the problems without much loss of energy and time. Once it is achieved, it is termed as troubleshooting, and becomes documentable.

1.2 Objectives and Approach

In refineries, capacity utilization, energy intensity index, process efficiency, and on-stream factor/availability are the key parameters that contribute a lot to refinery profitability and sustainability. Like other benchmarking organizations, Solomon Associates, USA, has developed all benchmarks with respect to the performance of refineries. The author has been associated with the audit of Indian refineries in benchmarking studies.

Petroleum Refineries: A Troubleshooting Guide
Dhananjoy Ghosh

ISBN 978-981-4877-81-7 (Hardcover), 978-1-003-18040-1 (eBook)
www.jennystanford.com

Various operational upsets in refineries result in poor gradation in refinery ranking with respect to all the above-mentioned parameters.

Hence, troubleshooting during any operational problem, when it arises, refinery operating crew get engaged to resolve the issue at the earliest to save operational losses as mentioned above.

For example, if the standby cannot be started in time due to the failure of a pump or a compressor, this leads to the shutdown of the process units or reduction in capacity; similar problems may also occur in distillation columns, vessels, reactors, exchangers, piping and fittings, which also cause a shutdown of the process unit or reduction in capacity. Intermittent failure on these accounts or delay in resolving the issues ultimately causes downward gradation of the above-mentioned performance parameters thereby resulting in heavy losses to the refinery.

Hence, having a good knowledge and expertise in troubleshooting reduces these losses. Though basic theoretical knowledge on the subjects is necessary for troubleshooting, experience plays a great role in quick troubleshooting as discussed earlier also. To facilitate quick troubleshooting, key issues in troubleshooting are discussed in respective chapters of the book. With the right background and experience along with apt for troubleshooting, reflect develops among the operating professionals; these qualities among the operating crew and the team leads facilitate troubleshooting very quickly minimizing the operational loss.

However, it is not wise to reinvent the wheel, i.e., troubleshooting experiences should be well documented and be made traceable in need to avoid the repetition of hit and miss trials when the same problem arises.

For continuous engagement without break in petroleum refineries for more than three decades, it was possible to know the ins and outs of the processes in association with almost all refinery processes in various forms from ground-level engineer in the field up to monitoring engineer for technical services and subsequently in a leadership position responsible for the whole refinery process units and offsite.

In documenting all the experiences, an effort has been made to explain in detail the problems first, and then troubleshooting approaches and solutions achieved have been narrated in all the cases as much as needed. Troubleshooting descriptions are mainly of text nature with reasoning and illustrations (wherever required); formulae, correlations have been used in the section where key issues in troubleshooting have been discussed. Hence, in discussing case studies, it is assumed that readers will be able to link the analogies in troubleshooting with mathematical correlations wherever required. Lessons on the respective troubleshooting cases have also been focused mostly. In some cases, where there were multiple causes or there were multiple probable solutions for a particular problem, and also where cases studies in those areas were enormous and commonly experienced, general demonstrations on troubleshooting are given on those scenarios.

So massive categories of troubleshooting experiences have been shared through this book that it will be easy for the engineers not only to become conversant to tackle those types of problems, if they arise, but also to arrive at quick solutions for non-repeating problems thereby enhancing their analytical skills and refinery profitability.

Troubleshooting skills of the engineers engaged in operation depends not only on their pre-education in alma mater and their engagement in operation of units, but also on knowledge in other disciplines like mechanical, electrical and instrument engineering. Hence, to build up knowledge on other disciplines as mentioned above, troubleshooting experiences with respect to problems in the processes due to causes contributed by reasons in multidiscipline engineering have been focused equipment-wise. Troubleshooting of almost on all technologies operative in petroleum refineries has also been discussed in separate chapters as follows.

Chapter 2

Traceability of Relevant Troubleshooting Experience

Traceability is one of the most discussed subjects in refineries which contribute a lot with respect to profitability and sustainability in refinery operation. Thrust is given on this aspect in ISO standards fulfilling criteria. All ISO accredited refineries, hence, give much importance to traceability of documents/knowledge.

While troubleshooting was concentrated on problems arising in a process unit's operation and in offsite areas operation in refineries, it was observed that problems are encountered in a particular furnace, column, exchanger, pump, compressor, other equipment/fittings, or in the process technology, which resulted in upset of the whole operation of the unit and the refineries solve the problems in due course.

Hence, to ensure easy traceability of the relevant experiences, gained in addressing the plant problems in concern, an effort has been made to list out all experiences relating to major types of equipment as mentioned above and also with respect to minor types of equipment like pressure safety valve (PSVs), isolation valves, Induced Draft/Forced Draft (ID/FD) fans, etc. In some specific case studies, it was found to be difficult to draw a distinction with respect to the identification of the root causes of problem whether it was due to column or exchanger; in those cases, the case study has been

Petroleum Refineries: A Troubleshooting Guide
Dhananjoy Ghosh

ISBN 978-981-4877-81-7 (Hardcover), 978-1-003-18040-1 (eBook)
www.jennystanford.com

repeated in both column and exchanger category and the repetition mentioned in the respective areas. Hence, readers may have to explore alternative categories of troubleshooting in the index for a particular case in extreme cases where it is found missing in an expected category.

Experiences in utilities and offsite areas have been documented under separate heading irrespective of categorizing against equipment.

There have been experiences in troubleshooting in process units where problems arose on poor performance with respect to technology in use/operation due to which product quality, yield, or unit capacity was getting limited. Troubleshooting experiences of this kind have been documented under a separate heading. All refinery process technologies have been covered in this category, including sulphur recovery unit (SRU) which is used for SO_2 emission control of refineries.

Like the traceability issues in documenting all troubleshooting experiences as discussed above, operating engineers also should ensure proper placement of the experiences in all respective process units and offsite either in distributed control system (DCS) screens and/or as hard copy documents in the respective work stations to ensure easy reach in need.

If possible, it is advisable to keep the book as a ready reckoner for individual engineers to refer with respect to traceability to colleagues in their relevant discussions.

Chapter 3

Methodologies in Troubleshooting

The first and foremost point is operating personnel should always be able to compare actual operating data with design operating data and more preferably with a test run operating data (when achievement of design performance is checked).

Following are the pre-requisites for good troubleshooting:

a. Logbook should be well maintained in operation.
b. Piping and instrumentation diagram (P&IDs) and process flow diagram (PFDs) should be kept updated.
c. Maintenance and inspection (M&I) shutdown activities should be well recorded.
d. All equipment datasheets should be available.
e. Maintenance history of rotaries should be well recorded.
f. In addition to having knowledge on respective disciplines, engineers should be apt to gather knowledge in petroleum technologies also.
g. Generally, chemical engineers are engaged with a responsibility to control the process operations with services from multi-disciplinary engineers. But, in a refinery, chemical engineers should also have the aptitude to gain knowledge in mechanical, electrical, instrument and civil engineering to some extent to facilitate understanding and communicating to all concerned.

Petroleum Refineries: A Troubleshooting Guide
Dhananjoy Ghosh

ISBN 978-981-4877-81-7 (Hardcover), 978-1-003-18040-1 (eBook)
www.jennystanford.com

h. Concept on HAZOP study should be learnt.

Similarly, the following procedures should be applied to troubleshoot the problems quickly:

a. Understand properly the unwarranted deviation which has occurred in the system.
b. To find whether a problem in any equipment was due to the process or not and if yes, what change has occurred in the process. If not, whether the problem should be due to an inherent problem that occurred in the equipment or due to ignoring some minor problem for long.

   Note: while thinking on the process, colleagues should be discussed for exchange of views, and for analysing on the types of equipment, engineer colleagues in respective disciplines, viz., mechanical/electrical/instrumentation/civil engineers should be engaged in the thought process.
c. For problems in product quality, yield and/or unit capacity limitation to maintain product qualities and product slate not attributed to any equipment, analysis on the technology in operation should be done thoroughly. To do it effectively, one should be well prepared with knowledge gained through training on that technology. Licensor's commitment to the process and/or catalyst(s) should be properly checked and should also check whether any deviation occurred or not in the process from licensor's guidelines.
d. Sometimes problems occur out of issues in metallurgy not considered in the design or engineering adequately; hence, for any corrosion problem in doubt, metallurgists of the organization should be engaged in the discussions.
e. Many problems arise in refineries due to malfunctioning or limiting of instruments and problems in a process control loop. So, while analysing the instruments or any control loop, instrument engineers should be engaged in troubleshooting.
f. Sometimes it is confusing to understand whether the problem is in electrical equipment and/or its circuit or not; hence, electrical engineers also should be involved in the troubleshooting discussion.
g. Multidisciplinary engineers can be called in the second phase of discussion after completing a primary review among the operating crew.

Chapter 4

Key Issues in Troubleshooting

4.1 Pump

Pump failure/malfunctioning are sometimes observed to be the root cause of the problem. A pump operation is upset with an abnormal flow rate of the liquid from the process which it can't handle as per its datasheet like very low flow rate below pump turndown capacity or due to very high flow than pump maximum capacity; both the cases result in level build up in the column/vessel from which it was taking suction; sometimes, its net positive suction head (NPSH) availability problem occurs, which can also result in similar effect-when available NPSH comes down to a level below its specified requirement in the datasheet, it loses suction and causes vapour locking inside pump casing which understood through knocking sound and/or fluctuation in the pump discharge pressure as seen from the discharge pressure gauge. It should be noted that the pump which is taking suction from a closed vessel/column, its available NPSH is not relevant to vapour pressure of the liquid as the vapour pressure of a liquid is the same as total pressure in the vessel/column from which it is taking suction; here, available NPSH changes only with the level of the liquid in that vessel/column which is positive head, and frictional pressure drop from vessel/column

Petroleum Refineries: A Troubleshooting Guide
Dhananjoy Ghosh

ISBN 978-981-4877-81-7 (Hardcover), 978-1-003-18040-1 (eBook)
www.jennystanford.com

bottom up to pump suction which being opposing the flow should be subtracted from the level head to get the available NPSH; hence even with a good level in the column/vessel, available NPSH may be lower if there is abnormal property change of the liquid or its temperature resulting in higher friction loss in the pump suction. In some specific cases, with everything normal in operating parameters, lower NPSH is realized which was found to be due to some changes in pump design like the introduction of an inducer in the pump suction nozzle which may not conform to the requirement. In atmospheric suction like when pump takes suction from an offsite storage tank, all parameters should be taken into account as per the formula of NPSH as follows unlike above as in that case the total pressure inside the tank would be atmospheric while there would be a different value for the vapour pressure of the liquid. Mathematical expression for NPSH with illustration is shown as follows:

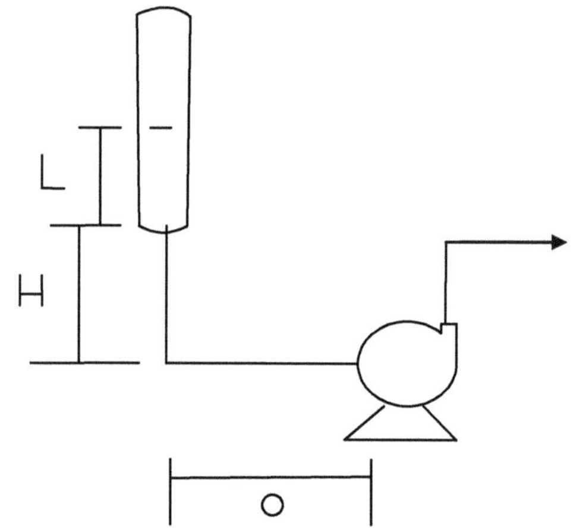

Figure 4.1 Flow dynamics to explain NPSH.

Figure 4.1 depicts the scenario of pumping a liquid from a closed column/vessel. Available NPSH to the pump is given by:

$$\begin{aligned} NPSH_{available} = {} & \text{Total pressure in the column/vessel} \\ & - \text{Vapour pressure of liquid} \\ & + \text{Gravity head at pump suction} \\ & - \text{Frictional pressure drop in suction line} \qquad (4.1a) \end{aligned}$$

As said earlier, the vapour pressure of the liquid is the same as the total pressure in the column/vessel; the final equation for available NPSH should be:

$$\begin{aligned} NPSH_{available} &= \text{Gravity head at pump suction} \\ &\quad - \text{Frictional pressure drop in suction line} \\ &= L + H - \text{Frictional pressure drop in suction line} \quad (4.1b) \end{aligned}$$

To calculate pressure drop, use Darcy's equation or refer to the monograph (Figure 4.2) as follows.

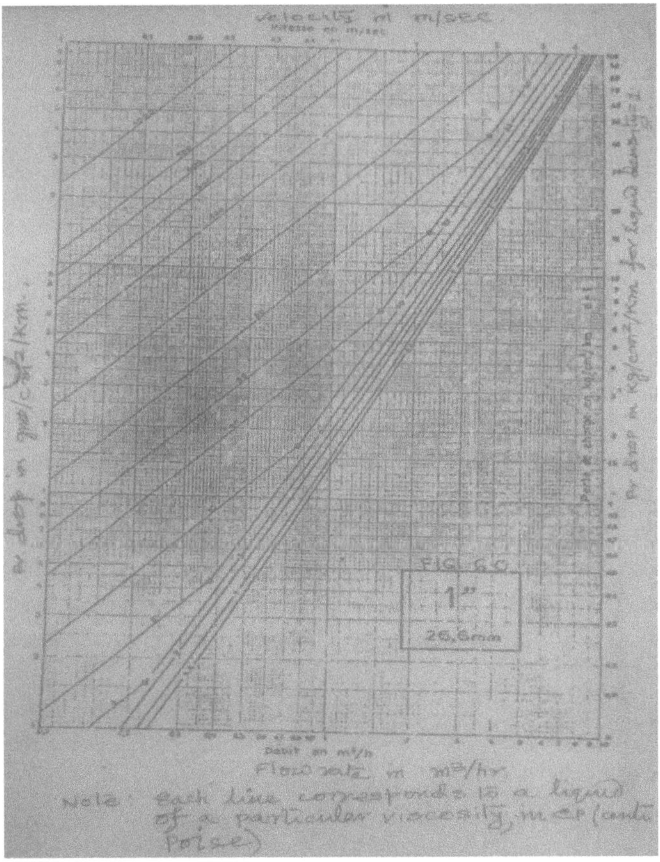

Figure 4.2 Relation between frictional pressure drop and volumetric flow rate/velocity for a fixed pipe diameter and fluid viscosity

The graph shown in Fig. 4.2 can be used for a line size of 1″ diameter. Similarly, there are graphs for all line sizes. In the *x*-axis of the graph, the bottom line figures represent flow rate in m^3/h and top line figures represent velocity in m/sec; hence, any one of the two can be used. In the *y*-axis, the RHS figures represent frictional pressure drop in $kg/cm^2/km$ for liquid density at 1 while LHS figures represent the same in $gms/cm^2/km$ for liquid density at 1. Each line corresponds to liquid of particular viscosity in centipoises (cp).

With the given inputs as mentioned above, once the frictional pressure drop is obtained from the *y*-axis, the same is to be multiplied with density of the liquid at pumping temperature to get the actual pressure drop in $kg/cm^2/km$ or $kg/cm^2/1000$ m; multiplying the same with the suction line length (in meter) of (H + O) as shown in the above sketch (Fig. 4.1), the total frictional pressure drop in pump suction line can be obtained.

To get the density of the liquid at pumping temperature, instead of using density from column/vessel datasheet, the following equation can be used which is applicable for petroleum hydrocarbon which is more realistic.

$$\text{Density}_{\text{at P.T.}} = \text{Density}_{\text{at }15°\text{C}} - 0.0006\,(\text{P.T.} - 15) \qquad (4.1c)$$

where P.T. is pumping temperature.

The gravity head of the pump suction is (L + H) to be taken in meter scale of measurement. If the liquid level is 40%, then equivalent length can be obtained from column/vessel data sheet taking the length of level from control tapings; also, H can be obtained from the datasheet where the height of the column/vessel from grade level is given.

The required NPSH of the pump as given in the pump data sheet is generally considered at 0% liquid level; hence, if the available NPSH at pump suction is to be compared with the required NPSH of the pump, then as a worst case, the available NPSH also should be calculated considering only the height of the column/vessel from ground level where the pump is installed. So, available NPSH simply becomes:

$$\text{NPSH}_{\text{available}} = H - \text{Frictional pressure drop at pump suction line;}$$

all in the meter.

If available NPSH is more than required NPSH, then there will be no cavitation, i.e., no vapour locking at pump suction; however, the pump will deliver desired flow steadily if its discharge line length, size, fitting, instruments, equipment located in the discharge lines are all as per design or on the other hand, the differential pressure across the pump discharge and its suction has been taken higher than the sum of all pressure drops in the discharge line as mentioned above.

Other criteria affecting pump performance

Change in operating temperature in addition to affecting available NPSH as mentioned above also affects the mechanical seal of the pump which is designed up to a max temperature thus making it fail. For seal protection, generally double seals with seal flushing/ circulation schemes are provided as per API standards as may be applicable for the service.

Sometimes, the internals of the pump found corroded which should be due to the wrong selection of metallurgy, but if it is found eroded either in the casing or in impeller, it can be due to vapour locking as a result of low NPSH or due to pumping corrosive liquid which in turn results in low available NPSH and/or lower capacity of the pump.

Sometimes, operating a single-stage pump with high differential pressure but at turndown capacity may result in high unbalancing thrust in the shaft causing erosion in the pump casing inside. Operating at turndown ratio for long, in any case, causes a temperature rise in the centrifugal pump as full rotational energy of the motor is not getting scope to be converted to translational energy due to pinching of pump discharge flow.

Issues with various categories of pumps

The above discussion was common for all types of pumps except the pump temperature rise issue which is valid only for the centrifugal pump as discussed above.

We know there are basically two types of pumps, viz., centrifugal and positive displacement pump; positive displacement pumps are also of different sub-categories, viz., reciprocating pump, screw pump, gear pump, lobe pump, etc. In refineries, mainly reciprocating pumps are in use, and in some cases with a very high viscous liquid, screw pumps are used, and many a time problems encountered with a reciprocating pump which is discussed as follows.

In a reciprocating pump, the following accessories are there:

- Loading/unloading valves are used both at suction and discharge ends of the pump cylinder.
- Pump casing/cylinder.
- Piston with piston rod and ring (disc) with lining.
- Safety valve at the discharge end.
- Packing glands or mechanical seal depending on the liquid used.
- Vibration reducer.
- Crankcase and stuffing box.

Loading/unloading valves are fitted with sitting springs which sometimes malfunction resulting in either no discharge pressure developing or low discharge flow; this problem also can occur with the damage of the valve seat(s) or malfunctioning of the instrument logic system when loading/unloading valves operations are guided by instruments impulse.

The safety valve is provided to avoid pump rupture by discharge pressure shoot up beyond pump's design pressure due to its discharge block valve stuck at the close condition or if there is a blockage in the discharge line.

Sometimes, it may happen that there is a partial blockage in the discharge line or some valve(s) in the discharge circuit is/are not fully open; in that case, there will be low flow from the pump.

Sometimes, there may be frequent gland packing/mechanical seal leakage which is due to either poor quality of gland packing/ mechanical seal procured or due to major variation of operating temperature or liquid viscosity or corrosiveness. Low flow may occur also due to damage of piston ring lining. Gland/mechanical seal leak may also result in abnormal vibration/oscillation of the pump—in that case, either it is due to vibration reducer malfunctioning (if used) or due to poor alignment of the pump. Poor alignment causes pump vibration in all types of pumps.

Failure in the crankcase generally is caused by poor lubrication while failure in the stuffing box is vulnerable in the case of reciprocating compressor discussed in the compressor section.

Issues with pump efficiency

To discuss various problems resulting in poor efficiency of the pump, let us refresh the knowledge on pump efficiency formula as follows:

$$\text{The overall efficiency of the pump, } e_f = e_{fm} \times e_{fp} \qquad (4.1d)$$

where e_{fm} and e_{fp} are efficiencies of pump drive and mechanical efficiency of the pump, respectively; for electric motor driven, drive efficiency is called electric efficiency.

$$e_{fm} = \text{BHP}/\text{IHP}$$

where BHP is the brake horsepower or output power of the motor, i.e., actual energy delivered by the motor to drive the pump which is given in the motor datasheet.

IHP is the indicated or input horsepower to the motor and calculated as follows:

$$\text{IHP} = \sqrt{3} \times I \times V \times \text{phase lag} \times \text{conversion factor from watt to horsepower}$$

where I = actual current drawn by the motor in ampere; V = motor voltage; phase lag is the power factor which should be 0.9 plus in an installation.

The value $\sqrt{3}$ is applicable for a three-phase motor which is as usual in a petroleum refinery.

Pump mechanical efficiency, e_{fp} can be obtained from the pump characteristic curve provided by the pump vendor for actual operating point, i.e., pump discharge flow rate and pump differential pressure (the difference between the pump discharge and suction pressure).

Evaluating the values of e_{fm} and e_{fp} as mentioned above and putting them in Eq. 4.1d, one can get the overall efficiency of the pump.

From above, it is evident that poor overall efficiency of the pump can be due to high current drawn by the motor as a result of poor motor insulation effectiveness, motor winding temperature rise or higher thrust required caused by motor bearing or pump bearing or higher pump differential pressure as compared to design.

4.2 Compressor

Before going to details of problems encountered in compressor operation, let us first discuss the principles in gas compression and various types of compressors used in the operation as follows:

As the liquid is generally incompressible, here compression means gas compression.

There are two types of compression, viz., isothermal compression and adiabatic compression.

Isothermal compression is the compression where pressure increases by compressing gas, i.e., reducing the gas volume at a constant temperature. Now, while compressing a real gas, its temperature obviously would rise, theoretically one can assume that if pressure is increased very slowly so that its temperature does not rise, then it is isothermal compression, but this takes infinite time to compress and thus it is not a practical process; hence, isothermal compression is not realistic. One may argue that temperature can be kept constant in a real case by putting a cold jacket in the compressor body, but that is not valid as here an external factor, i.e., the heat generated in the system is being taken out by a third medium like jacket cooling. Work done by the compressor in isothermal compression is given by:

$$W = -P\int dV \quad (4.2a)$$

$$= -nRT \times (1/V)\int dV\text{; due to the constant temperature process} \quad (4.2b)$$

$$= -nRT \times \ln(V_f/V_i)$$

$$= -2.303 \times nRT \times \log(V_f/V_i)$$

$$= 2.303 \times nRT \times \log(V_i/V_f) \quad (4.2c)$$

$$= 2.303 \times nRT \times \log(P_f/P_i)\text{; due to } P_fV_f = P_iV_i \quad (4.2d)$$

where R is gas constant, n is the moles of gas flow rate, T is the constant temperature, V_i and V_f are initial and final volume, respectively, and P_i and P_f are initial and final pressure, respectively.

Similarly, the work done formula for adiabatic compression can be calculated. For the adiabatic process, the following PV equation is valid due to change in temperature thus causing a change in internal energy and given by:

$$PV^{\gamma} = \text{Constant, say } K \quad (4.2e)$$

where P = pressure of the gas; V = volumetric flow rate of the gas; γ = ratio of specific heat at constant pressure and at constant volume. The value of γ is 1.66 for mono-atomic gas, 1.4 for poly-atomic gas; in petroleum refinery for hydrocarbon it is usually assumed at 1.3 Similarly,

$$P_f V_f^{\gamma} = P_i V_i^{\gamma} = K \quad (4.2f)$$

where P_f and V_f are final or discharge pressure and volumetric flow rate at discharge, respectively, and P_i and V_i are initial or suction pressure and volumetric flow rate at suction, respectively.

Now, work done by the compressor is given by:

$$W = -P\int dV$$
$$= -K \times (1/V^{\gamma})\int dV\text{; using Eq. 4.2e}$$
$$= -K \times \int V^{-\gamma} dV$$
$$= -K/(1-\gamma) \times (1/V_f^{\gamma-1} - 1/V_i^{\gamma-1})$$
$$= -1/(1-\gamma) \times (K/V_f^{\gamma-1} - K/V_i^{\gamma-1})$$
$$= -1/(1-\gamma) \times (P_f V_f^{\gamma}/V_f^{\gamma-1} - P_i V_i^{\gamma}/V_i^{\gamma-1})\text{; using Eq. 4.2f}$$
$$= -1/(1-\gamma) \times (P_f V_f - P_i V_i)$$
$$= -P_i V_i/(1-\gamma) \times (P_f V_f/P_i V_i - 1)$$
$$= -P_i V_i/(1-\gamma) \times ((P_f/P_i)(V_f/V_i) - 1))$$
$$= -P_i V_i/(1-\gamma) \times ((P_f/P_i)(P_i/P_f)^{1/\gamma} - 1))$$
$$= -P_i V_i/(1-\gamma) \times (P_f^{1-1/\gamma}/ P_i^{1-1/\gamma} - 1)$$
$$= -P_i V_i/(1-\gamma) \times ((P_f/P_i)^{1-1/\gamma} - 1))$$
$$= P_i V_i/(\gamma-1) \times ((P_f/P_i)^{1-1/\gamma} - 1)) \quad (4.2g)$$

where, P_i and V_i are suction pressure and volumetric flow rate at suction temperature, and P_f is the discharge pressure of the compressor.

Considering the gas being an ideal gas, Eq. 4.2f can also be converted to the following:

$$P_f^{1-\gamma} T_f^{\gamma} = P_i^{1-\gamma} T_i^{\gamma} \quad (4.2h)$$

where T_i and T_f are the suction and discharge temperature of the compressor; thus discharge temperature can be calculated given suction temperature, suction pressure and discharge pressure.

Like pumps, there are characteristic curves for compressors also along with a complete data sheet including motor or turbine whatever drive is used. Hence, like above shown in the case of pump, similar efficiency calculation can also be done for the compressor, and if the efficiency is found low it may be due to higher current drawn by the motor or higher steam consumption by the turbo-drive or due to higher back pressure in the processing system. In adiabatic compression as temperature rises during compression, compressor suction temperature should be kept lower to prevent discharge temperature increases beyond the allowable limit. Hence, when the differential pressure requirement is high, in that case, compression is done in multi-stages with an external cooler in between the stages.

Issues with various types of compressor

In a centrifugal compressor, their mechanical integrity is generally well established for three/four decades, still, sometimes minor damage in impellers is observed and happens very rarely. Its main operational problem occurs due to improper seal oil and lube oil circuit functioning; however, to protect failure against this, trip mechanisms are provided. If time to time, liquid carry-over occurs in the compressor as a result of upstream suction vessel demister pad damage or upset in the process beyond control by demister, then its capacity decreases over a period of time reflected by an increase in compressor discharge temperature though the system back pressure did not increase; the liquid carryover causes gradual damage to labyrinth seal placed in between stages of the compressor which in turn effect in some leakage of gas from high-pressure end to low-pressure which being hotter increases suction pressure and temperature of gas with ultimate effect on lowering discharge pressure and higher discharge temperature. Otherwise, it is well protected by various safety interlocks.

The compressor is protected against vibration with the provision of thrust bearing at compressor both ends and with journal bearing at both ends of the drive-in case of the heavy-duty machine. There is a voting mechanism in the trip system with these bearings, and it happens sometimes that there is a spurious trip due to the malfunctioning of the impulse system to the bearings. Similarly, there are multiple impulse lines against motor winding temperature which also cause spurious trips. Interlock system malfunctioning in lube oil and seal oil system also causes machine trip, but non-start of standby lube oil/seal oil pump happens if turbo-driven is kept as stand by; hence, turbo driven lube oil/seal oil pump should be run as main pump and motor-driven pump should be used as standby because the motor-driven pump can instantaneously take full speed when started unlike turbo driven pump. In case the main drive is turbo driven, then sometimes it trips due to malfunctioning of mechanical trip system tripping the lever at the steam line on the higher speed of the drive; hence, the electronic trip system also should be used and interlock logic should be maintained such that unless electronic trip device shows the higher speed of the drive, the machine should not trip.

For the reciprocating compressors, adequate safety interlocks were not available earlier, but nowadays, different on-line protection systems are implemented at least to get predictions of future failure which helps a lot in saving the equipment from unplanned/ catastrophic failure.

General problems of reciprocating compressors are like rising in discharge temperature due to poorly designed loader valves or due to damage of loader valves as a result of liquid carryover or due to restriction in the flow path in the downstream, as a safety measure against low flow and high back pressure, anti-surge protection system is provided for big compressor; for low capacity compressor, its safety valve at discharge takes care. Low flow also occurs if gas becomes lighter, viz., of low molecular weight in the composition. If gas becomes heavier, then compressor volumetric flow will not be a problem but the turbine/motor drive side will be a problem for desired volumetric flow.

Another problem like no rise in discharge pressure is due to malfunctioning of loading mechanism, for e.g., in loading/unloading mechanism, suction loader valve should be open during compressing cycle and should be closed after compression is complete;

simultaneously, discharge loader valve should remain closed during compression and should open after compression is complete; all these sequencing operations should be in auto mode as per setting.

For the multistage compressors, integrity between stages like inter condensers functioning and balancing line integrity are important to look into.

The similar problems with the trip system in compressor side, drive side and in auxiliary system like lube oil/seal circuits also occur here like centrifugal compressor as mentioned above; hence, the philosophy of operation and care in instrumentation should be followed here also like above.

4.3 Heat Exchanger

Before discussing various problems arising in heat exchangers operation, its working principles are highlighted as follows:

In heat exchangers, heat released by a hot fluid is considered the same as heat absorbed by cold fluid assuming that heat insulation on the shell side, i.e., external side of the exchanger is well insulated to avoid heat loss to the ambient. Thus heat equations are valid as followed:

$$Q = M \times C_{\mathrm{p}} \times (T_1 - T_2) = m \times c_{\mathrm{p}} \times (t_2 - t_1) \tag{4.3a}$$

where M and m are the mass flow rate of the hot and cold fluid, respectively and C_{p} and c_{p} are the specific heat of hot and cold fluid, respectively; the suffix 'p' is used with specific heat to denote that the same is at constant pressure if the fluid is gas instead of liquid. The temperature of hot fluid and cold fluid are designated as shown below:

Hot fluid $T_1 \rightarrow T_2$

Cold fluid $t_2 \leftarrow t_1$

It is seen from above that the flows are counter current as counter-current flows ensure a higher temperature gradient across the full length of the exchanger unlike co-current flows.

With respect to the heat exchange efficiency of the exchanger, there is another kind of heat duty equation as follows:

$$Q = U_d \times A \times \text{LMTD} \quad (4.3b)$$

where U_d is the overall heat transfer coefficient between hot and cold fluid, A is the heat transfer area as provided in the exchanger, and LMTD is logarithmic mean temperature difference between hot and cold fluid as explained below:

$$\text{LMTD} = (\Delta T - \Delta t) \,/\, \ln(\Delta T - \Delta t) \quad (4.3c)$$

where $\Delta T = T_1 - t_2$ and $\Delta t = T_2 - t_1$. It may be noted that instead of LMTD, average temperature gradient like $(\Delta T + \Delta t)/2$, i.e., $(T_1 + T_2)/2 - (t_1 + t_2)/2$ can be used in above heat transfer equation, but LMTD is more representative of temperature gradient between hot and cold fluid.

The overall heat transfer coefficient, U_d is a combination of heat transfer coefficients of both hot and cold fluid along with fouling factor of both these two streams when flows through a heat exchanger are based on the principle that overall resistance to heat transfer is a sum of heat transfer coefficient of hot and cold fluid plus fouling (dirt) resistance both in the tube side and shell side of the heat exchanger through which respective fluid flows, and that heat transfer resistance is inverse of heat transfer coefficient. The equation is shown as follows:

$$1/U_d = 1/h_t + 1/h_s \times (-A_o/A_i) + f_t + f_s \times (A_o/A_i) \quad (4.3d)$$

where h_t is the heat transfer coefficient in the tube side and h_s is the heat transfer coefficient in the shell side of the exchanger through which respective fluid flows; A_o is the outside surface area of all tubes together based on tube OD (outside diameter) and A_i is the inside surface area of all tubes together based on tube ID (inside diameter); f_t and f_s are the fouling resistance in the tube side and shell side respectively.

As the heat transfer coefficient is expressed per unit surface area, all values of the respective heat transfer coefficient should be expressed in terms of the same surface. Hence, to express U_d in terms of the same surface area, i.e., either A_s or A_t, say in terms A_t, area correction factors have been used as shown in Eq. 4.3d above.

The heat transfer coefficient, h_s and h_t in Eq. 4.3d are the clean transfer coefficients in the shell side and tube side respectively, and their values are determined by using the Dittus-Boelter equation and convection heat transfer formula using the given properties and operating conditions of the hot and cold fluid in the tube side and shell side respectively or vice versa as decided during the design of the heat exchanger. The values now a days can be evaluated using a computer program quickly instead of manual use of this Dittus-Boelter equation taking a longer time.

If it is not the requirement to design a heat exchanger, but to monitor the performance of existing exchanger in operation, then there is no point in calculating h_t and h_s as mentioned above; instead in operation monitoring, one should know whether the exchanger is performing as per design after commissioning and achieving design operating conditions or not; also, it is necessary to know whether exchanger performance is deteriorating during the progress of operation over a period of time and decide on the outage of the exchanger for cleaning. For this purpose, Eq. 4.3b should be used to find out the value of U_d from design data on temperature, Q and A as available in the heat exchanger data sheet; sometimes, however, data sheets directly mention the value of U_d. Same U_d can also be evaluated using actual operating data of the exchanger using Eq. 4.3b. To compare design U_d and actual U_d, either operating data like flow rate and inlet temperatures of hot and cold fluids should be maintained as the same in the design along with their physicochemical properties to be the same in the design. If physicochemical properties cannot be achieved, then the designer should be asked to check the value of U_d with the actual properties of the fluids to compare the actual performance with the design. But if the objective is only to know the fouling of the exchanger over a period of time since commissioning or since startup after an exchanger cleaning job, then simply evaluate the value of U_d during that time and compare with design U_d to get the difference to conclude on the requirement of an outage of the heat exchanger for cleaning.

Issues with heat exchangers' performance

Heat exchangers' major problem arises due to pre-mature fouling on either side. If shell side fouls, it must be due to congealing of liquid or coke build up beyond expectation; in that case either root-cause at upstream should be eliminated or in extreme cases, switch over of fluid service between tube and shell side should be explored. Tube side fouling occurs due to excessive viscosity of liquid reducing velocity and subsequent choking; in that case, viscosity reducer like inserts should be provided in the tubes or exchanger should be modified from shell and tube type to Helical tubes exchanger.

Sometimes, it is found in the case of a reboiler that heat transfer decreased though operating parameters are normal as well as there is no dirt found inside or outside the tube surface; just after replacing the tube bundle, and it is found that the problem got solved. Hence, heat transfer fouling resistance can also be due to formation of heat resistant film on the inside or outside surface of the tubes due to continuous operation over a period of time or using a tube bundle with already formation of invisible scale on the tubes.

In cooler, generally, the tube inside gets fouled and sometimes choked due to high-temperature rise of water in tube side which causes a deposit of salt present in water; improved water treatment in a cooling tower, the higher velocity of water in tube side, increasing water flow help to control this fouling.

Tube leakage also occurs in choked tubes due to stagnant water inside tubes and due to no heat transfer. In vapour coolers also, tube leaks occur but it is seen due to stress corrosion cracking caused by undesired impurities in the gas in the shell side or improper metallurgy of tubes selected. Tube leakages in exchangers are uncommon and leak in this mostly found to be due to leakages in floating head gasket between tube and shell side.

Thermal shock in heat exchangers causes safety hazard by leakage of gas/liquid to outside and also sometimes damage exchanger internals and tubes to create malfunctioning and tube leakage problems subsequently.

Issues with various types of heat exchangers

So far, in the above only one type of heat exchanger, viz., shell and tube type has been mentioned though all above are applicable for all types of a heat exchanger. However, it is necessary to know about various types of heat exchangers having specific advantages/disadvantages of one type over the other through the above principles are the same.

Petroleum refineries also use different types of heat exchangers in different areas as explained as follows:

- Shell and tubes
- Helical tubes and shell
- Shell and tubes with inserts inside tubes
- U-tube
- Air-Fin cooler
- Plate type exchanger

Shell and tube type exchangers are most common not only in refinery and petrochemical industries but also in process chemical industries.

Shell and tube type exchangers with variations like helical tubes or inserts within tubes are used for using viscous fluid inside the tubes to get the turbulence for getting a higher heat transfer coefficient.

Shell and tube type exchangers can be cleaned offline to get back the heat transfer efficiency but the shell and tube type exchangers are to be provided with some excess heat transfer area as there would be some unutilized tube surface even after providing a good design of baffles in the shell side; the baffles in the shell side not only increase the turbulence in the shell side but also helps in reducing the void space to some extent.

When a large heat transfer area is required in a single exchanger to save cost, then to avoid much loss of heat transfer surface, plate type exchangers are used but the heat transfer fluids should be clean as there is no scope of cleaning the plate exchangers.

Air-Fin coolers are used for cooling vapours when the quantity is large and there is a shortage of cooling water; in that case, vapours are first cooled in Air-Fin coolers followed by cooling in a water cooler where water flows through the tube side.

U tube exchangers are very simple and small in size and are used for very small heat duty cases which are sometimes required in the industries.

When Reynolds number is a criterion to get higher heat transfer coefficient and hence objective is placed to decide whether hot fluid or cold fluid should flow through tubes to get higher Reynolds number, dirty fluid claims to be in the tube sides so that exchangers can be cleaned in need; hence, many a time it is seen that refineries have modified shell and tube service depending upon pump pressure availability.

Air-Fin coolers, many a time get fouled up at external fins or sometimes air-fins get damaged; hence, in turn around shut down of the plant, these fins should be cleaned and maintenance care should be taken to avoid damage to air-fins.

To optimize cost depending upon operating conditions, the thickness of plates in plate type exchangers kept low; in that case, design care is taken to ensure no pressure surge across the plates during an emergency shutdown. Also, as the exchangers cannot be cleaned, not only clean fluid is selected for operation, but also strainer trap is used at the inlet of the exchanger; with high delta pressure, the unit should get tripped to avoid pressure differential across the plate if not desired as per design.

Metallurgy issues in exchangers

Sometimes, wrong metallurgy is found to be selected and used in the tube bundles; for e.g., in a water cooler, carbon steel tubes were selected where process fluid contained H_2S and the fluid was to be cooled from a high temperature of about 225°C to about 50°C with help of cooling water inside the tubes. The corrosive liquid at high temperature caused stress corrosion cracking resulting in rupture of the tube bundle. After replacing the tube bundle with cupronickel metallurgy, the problem got solved. While using chemicals, the metallurgy issues are generally found to be taken care of well but in the case of hydrocarbon processing, there have been many mismatches on metallurgy observed like above mentioned.

4.4 Furnace

Like earlier, here also, let us brush up on the theoretical aspect of fuel firing inside a furnace. Furnace whether box type or cylindrical type, whether vertical or horizontal firing, it has two sections, viz., radiation heat transfer section and convection heat transfer section. Also, when fuel is fired, it generates hot flue gas along with flame inside the radiation section; the flue gas after transferring heat from the flame through radiation to the furnace tubes, exits the radiation zone and enters the convection zone to further heat transfer to process fluid inside the tubes in the convection section. Even after the convection section, if there is enough heat in the flue gas, then water or saturated steam tubes are placed above the convection zone as economizer tubes or superheater tubes respectively to utilize the remaining heat in the flue gas before the flue gas exits from the stack to atmosphere. However, to maximize heat utilization, if the flue gas flow rate is adequately higher, then its heat is further utilized to heat up combustion air in the air preheater provided outside the furnace.

High tube metal or skin temperature in the radiation section is a common problem in refineries. There can be two reasons for higher skin temperature, viz., faulty skin temperature indication by the instrument thermocouple(s) or heat flux in the radiation section is really high. Hence, in the same firing zone in the furnace, there should be more than one indication and, also with a higher heat load for high capacity furnace operating with many fuel burners and many tubes, there should be skin thermocouples in the different firing zones. However, before doubting skin temperatures indication, one should understand whether actual heat load is high or not and also whether a flame pattern is unsteady like sweeping the tubes where skin temperature thermocouples are located; firing should be adjusted in the later case, but theoretical knowledge to be earned to solve the problem in the former case. Hence, the formula for radiation heat transfer is explained as follows:

$$Q_R = \sigma \times € \times A \times (T^4 - t^4) \qquad (4.4a)$$

where Q_R is the radiation heat transfer rate; σ is the Stiffen–Boltzmann constant; € is the overall emissivity of the hot flue gas/flame and furnace tubes surface considering both as gray body surface; A is the surface area of the tubes receiving radiation heat; T is the effective gas temperature/flame temperature generated out of fuel firing; t is the tube wall or skin temperature. Considering both flame and furnace tubes are in parallel, the overall emissivity, € is given by:

$$€ = 1/(1/€_R + 1/€_A - 1) \qquad (4.4b)$$

where $€_R$ is the emissivity of radiating flame/flue gas, and $€_A$ is the emissivity of furnace tubes.

Now, absorbed heat duty in the radiation section can be calculated by calculating the total absorbed heat duty of the process fluid (using process fluid inlet and outlet temperature, flow rate, specific heat, percent vaporization of process fluid and heat of vaporization to find sensible heat gain plus the heat of vaporization) and subtracting convection heat transfer (using convection inlet and outlet temperature of the process fluid and its flow rate and specific heat to get sensible heat gain in convection zone). Radiant heat absorbed by the process fluid can also be calculated directly using the radiation inlet and outlet temperature of the process fluid.

Similarly, radiation heat loss also to be calculated using a similar equation like Eq. 4.4a as given by:

$$Q_{RL} = \sigma \times €_O \times A_{FW} \times (T_O{}^4 - t_A{}^4) \quad (4.4c)$$

where Q_{RL} is radiation heat loss by the furnace outside wall to ambient; $€_O$ is overall emissivity of furnace outside wall and ambient, T_O is furnace outside wall temperature, t_A is ambient temperature, and A_{FW} is furnace wall surface area.

Hence, Q_R is also given by:

$$Q_R = Q_{AR} + Q_{RL} \quad (4.4d)$$

where Q_{AR} is radiant heat absorbed as explained earlier.

Various co-relations have also been developed to relate only with effective flue gas temperature/flame temperature, T by different authors; one such is given as follows:

$$F(T) = C \times T^4 + D \times T - B$$

or

$$\mathrm{d}F(T)/\mathrm{d}T = 4 \times C \times T^3 + D \quad (4.4e)$$

where $F(T)$ is the heat released by the fuel firing/combustion process, T is effective flue gas temperature or flame temperature in the radiation zone, and C, D, B are constants.

The above Eq. 4.4e should be solved using a computer program and adequate operating data are required to get the solution. The value of T next can be used in Eq. 4.4a above to know the value of t, i.e., tube wall or skin temperature. This value of t can next be used to calculate coke thickness built up inside the furnace tube by using conduction plus convection heat transfer formulae inside furnace tubes.

A case study in excel carried out for the same and relevant input data both with fuel oil (FO) and fuel gas (FG) firing used are shown as follows along with calculation for radiant heat loss, Q_{RL} as illustration:

Table 4.1 Excel work up to evaluate furnace tube skin temperature and furnace efficiency

INPUT		FO FIRING				
AMBIENT TEMPERATURE				60	°F	(FIXED)
AIR TEMPERATURE				81	°F	(VARIABLE)
FLUE GAS TEMPERATURE TO THE STACK				298	°F	(VARIABLE)
FUEL OIL TEMPERATURE				194	°F	(VARIABLE)
RELATIVE HUMIDITY				64	%	(VARIABLE)
FLUE GAS ANALYSIS	O_2			3	VOL%	(VARIABLE)
RADIATION HEAT LOSS				1.50	% OF LHV	(VARIABLE)
VERTICAL FLAT SURFACE						
SURFACE AREA				5366	FT^2	(FIXED)
SURFACE TEMPERATURE				111	°F	(VARIABLE)
HORIZONTAL FLAT SURFACE						
SURFACE AREA				933	FT^2	(FIXED)
SURFACE TEMPERATURE				116	°F	(VARIABLE)
FUEL GRAVITY				20	° API	(VARIABLE)
FUEL OIL ANALYSIS						
CARBON–HYDROGEN RATIO				7.36		(VARIABLE)
HIGHER HEATING VALUE		10562	KCAL/KG	19011	BTU/LB OF FUEL	(FIXED)
SULPHUR CONTENT				0.71	%	(VARIABLE)

INERT CONTENT					0.09	%	(VARIABLE)
TEMPERATURE OF ATOMIZING STEAM					423	°F	(VARIABLE)
PRESSURE OF ATOMIZING STEAM					110	PSIG	(VARIABLE)
QUANTITY OF ATOMIZING STEAM					0.3051	LB/LB OF FUEL	(VARIABLE)
LHV			9884	KCAL/KG	17806	BTU/LB OF FUEL	(FIXED)
FUEL							
FO	38	TPD	78	%	3480	LB/HR	(VARIABLE)
FG	10	TPD	22	%	963	LB/HR	(VARIABLE)
TOTAL	48	TPD	100	%	4443	LB/HR	
WET BULB TEMPERATURE			19	°C	66	°F	
DRY BULB TEMPERATURE			23	°C	73	°F	

DERIVED DATA				
C_p air	60	°F	0.24	BTU/LB°F
C_p fuel			0.48	BTU/LB°F
Enthalpy of Atomizing Steam	423	°F	1203	BTU/LB OF FUEL
Enthalpy of Atomizing Steam	60	°F	1088	BTU/LB OF FUEL
Relative Humidity (Using W/D Bulb Temperature)			64	%
qvapour	60	°F	0.5068	PSIA

(Continued)

Table 4.1 (*Continued*)

INPUT DATA FURNACE (GAS FIRED)					
INPUT REQUIRED			FG FIRING		
1	AMBIENT AIR TEMPERATURE		60	°F	(FIXED)
2	AIR TEMPERATURE		81	°F	(VARIABLE)
3	FLUE GAS TEMPERATURE TO THE STACK		298	°F	(VARIABLE)
4	FUEL GAS TEMPERATURE		140	°F	(VARIABLE)
5	RELATIVE HUMIDITY		64	%	(VARIABLE)
6	FLUE GAS ANALYSIS	O_2	3	VOL%	(VARIABLE)
7	RADIATION HEAT LOSS		**1.50**	% OF LHV	(VARIABLE)
	VERTICAL FLAT SURFACE				
	SURFACE AREA		5366	FT^2	(FIXED)
	SURFACE TEMPERATURE		111	°F	(VARIABLE)
	HORIZONTAL FLAT SURFACE				
	SURFACE AREA		933	FT^2	(FIXED)
	SURFACE TEMPERATURE		116	°F	(VARIABLE)
8	FUEL GRAVITY		0	° API	
9	FUEL OIL ANALYSIS				
	CARBON–HYDROGEN RATIO		0		
	HIGHER HEATING VALUE		0	BTU/LB OF FUEL	(FIXED/VARIABLE)

	SULPHUR CONTENT				0	%	
	INERT CONTENT				0	%	
10	TEMPERATURE OF ATOMIZING STEAM				0	°F	
11	PRESSURE OF ATOMIZING STEAM				0	PSIG	
12	QUANTITY OF ATOMIZING STEAM				0	LB/LB OF FUEL	
13	LOWER HEATING VALUE				20402	BTU/LB OF FUEL	
14	FUEL	(TPD)					(FIXED/VARIABLE)
	FO	38	78	%	3480	LB/HR	(VARIABLE)
	FG	10	22	%	963	LB/HR	(VARIABLE)
	TOTAL	48	100	%	4443	LB/HR	
15	WET BULB TEMPERATURE		18	°C	64	°F	
16	DRY BULB TEMPERATURE		17	°C	63	°F	

DERIVED DATA

C_p air	60	°F	0.24	BTU/LB°F
C_p fuel			0.53	BTU/LB°F
Relative Humidity(Using Wet /Dry Bulb Temp.)			64	%
qvapour	60	°F	0.5068	PSIA

(*Continued*)

Table 4.1 (*Continued*)

Radiation Loss with FO firing:

RADIATION LOSS				
Hr =	0.1548 × (*e*/0.9) × (((Ts/100)^4) – ((Ta/100)^4)))/(ts – ta))			BTU/HRSQFT°F
	1.071			BTU/HRSQFT°F
Qr =	hr × Area(A)× (ts – ta)			BTU/HR
	220130			BTU/HR
	63			BTU/LB
Where				
	e =	coefficient of radiation		
	Ts	°R	Temperature of hot surface	573
	Ta	°R	Ambient Temperature	541
	Ts	°F	Temperature of hot surface	113
	Ta	°F	Ambient Temperature	81
	Qr	BTU/HR		
	Hr	coeff. of conv. and radiation heat transfer		

Radiation loss with FG firing:

RADIATION LOSS				
Hr =	0.1548 × (e/0.9) × (((Ts/100)^4) – ((Ta/100)^4)))/(ts – ta))			BTU/HRSQFT°F
	1.071			BTU/HRSQFT°F
Qr=	hr × Area (A) × (ts – ta)			BTU/HR
	220130			BTU/HR
	229			BTU/LB
Where				
	e =	coefficient of radiation		
	Ts	°R	Temperature of hot surface	573
	Ta	°R	Ambient Temperature	541
	Ts	°F	Temperature of hot surface	113
	Ta	°F	Ambient Temperature	81
	Qr	BTU/HR		
	Hr	coeff. of conv. and radiation heat transfer		

(Continued)

Table 4.1 (*Continued*)

Evaluation of furnace thermal efficiency

Using the above input data, combustion worksheets, fuels heat values, i.e., LHVs, relative humidity of ambient, furnace excess air, furnace stack loss, and efficiency can be calculated. Hence, the balance data sheets as required and the outputs are given as follows:

Combustion worksheet with FO firing

FUEL COMPONENT	col 1	COL 2	COL 3	COL 4	COL 5	COL 6	COL 7	COL 8	COL 9	COL 10	COL 11	COL 12	COL 13
			1X2		3X4		3X6		3X8		3X10	N_2	3X12
	Vol	MW	TOT.WT.	NET HV	HEATING	AIR	AIR	CO_2	CO_2	H_2O	H_2O	LB N_2/LB	N_2
	Frac				VALUE	REQUIRED	REQUIRED	LB CO_2/ LB	LB	LB H_2O/ LB	LB		LB
			LB	BTU/LB	BTU	LB AIR/LB	LB						
CARBON		12	0.8733		0	11.51	10.052	3.66	3.196		0.000	8.85	7.729
HYDROGEN		2	0.1187	51600	6123	34.29	4.069		0.000	8.94	1.061	26.36	3.128
OXYGEN		32	0.0000		0	-4.32	0.000		0.000		0.000	-3.32	0.000
NITROGEN		28	0		0		0.000		0.000		0.000	1	0.000
CARBON MONOOIXDE		28	0	4345	0	2.47	0.000	1.57	0.000		0.000	1.9	0.000

FUEL COMPONENT	col 1	COL 2	COL 3	COL 4	COL 5	COL 6	COL 7	COL 8	COL 9	COL 10	COL 11	COL 12	COL 13
CARBON DI OXIDE		44	0		0		0.000	1.00	0.000		0.000		0.000
METHANE		16	0	21500	0	17.24	0.000	2.74	0.000	2.25	0.000	13.25	0.000
ETHANE		30	0	20420	0	16.09	0.000	2.93	0.000	1.8	0.000	12.37	0.000
ETHYLENE		28	0	20290	0	14.79	0.000	3.14	0.000	1.28	0.000	11.36	0.000
PROPANE		44	0	19930	0	15.68	0.000	2.99	0.000	1.63	0.000	12.05	0.000
PROPYLENE		42	0	19690	0	14.79	0.000	3.14	0.000	1.28	0.000	11.36	0.000
BUTANE		58	0	19670	0	15.46	0.000	3.03	0.000	1.55	0.000	11.88	0.000
BUTYLENE		56	0	19420	0	14.79	0.000	3.14	0.000	1.28	0.000	11.36	0.000
PENTANE		72	0	19500	0	15.33	0.000	3.05	0.000	1.5	0.000	11.78	0.000
HEXANE		86	0	19390	0	15.24	0.000	3.06	0.000	1.46	0.000	11.71	0.000
SULPHUR		32	0.0071		0	4.31	0.031	2.00	0.014		0.000	3.31	0.024
HYDROGEN SULPHIDE		34		6550	0	6.08	0.000	1.88	0.000	0.53	0.000	4.68	0.000
INERTS			0.0009										
TOTAL			1.0000		6123		14.152		3.211		1.061		10.880
PER POUND OF FUEL			1.0000		6123		14.152		3.211		1.061		10.880

Note: Similar combustion worksheet to be generated considering only fuel gas (FG) firing.

(*Continued*)

Table 4.1 (*Continued*)

Relative humidity with FO and FG firing

FO: ATOMIZING STEAM RATIO : 0.3051 LB/LB OF FUEL (ASSUMED OR MEASURED)

CORRECTION FOR RELATIVE HUMIDITY

MOISTURE IN AIR = $\frac{\text{qvapour} \times \text{Relative Humidity} \times 18}{14.696 \times 100 \times 28.85}$

0.014 LB OF MOISTURE/LB OF AIR

Where qvapour = vapour pressure of water at the ambient temperature (psia) – 0.5068 from steam table

LB OF WET AIR /LB OF FUEL REQUIRED = $\frac{\text{AIR REQUIRED}}{1 - \text{MOISTURE IN AIR}}$ =14.349

LB OF MOISTURE / LB OF FUEL = LB OF WET AIR /LB OF FUEL – AIR RQD

=0.198

LB OF H_2O_TOTAL /LB OF FUEL = HO FORMED +LB OF MOISTURE /LB OF FUEL + ATOMIZING STEAM

=1.564

These total water and moisture values are used in the excess air calculation sheet as follows.

Note: Similarly, relative humidity of air to be calculated considering only fuel gas (FG) firing.

Stack loss with FG as fuel

EXIT FLUE GAS TEMPERATURE				
	Te	=	388	°F

COMPONENT	COLUMN 1 LB OF COMPONENT FORMED / LB OF FUEL	COLUMN 2 ENTHALPY AT T BTU/LB FORMED	COLUMN 3 HEAT CONTENT BTU/LB OF FUEL (1 × 2)
CO_2	2.932	50	146.58
H_2O	1.818	100	181.80
N_2	12.346	58	716.07
AIR	4.788	55	263.34
TOTAL	21.883		1307.78

STACK HEAT LOSS Qs = 1307.78 BTU/LB OF FUEL

(Continued)

Table 4.1 *(Continued)*

Stack loss with FO as fuel

EXIT FLUE GAS TEMPERATURE			
	Te =	298	°F
COMPONENT	COLUMN 1 LB OF COMPONENT FORMED / LB OF FUEL	COLUMN 2 ENTHALPY AT T BTU/LB FORMED	COLUMN 3 HEAT CONTENT BTU/LB OF FUEL (1 × 2)
CO_2	3.2106233	50	160.53
H_2O	1.60	100	159.96
N_2	10.880443	58	631.07
AIR	2.586	55	142.25
TOTAL	18.277		1093.81
	STACK HEAT LOSS	Qs =	1093.81 BTU/LB OF FUEL

Note: In the above calculation, the amounts of CO_2 and N_2 are taken from the combustion worksheet as mentioned, while H_2O and air should be taken from excess air evaluation in the furnace.

CORRECTION FOR EXCESS AIR:

LB OF EXCESS AIR /LB OF FUEL

= $((28.85 \times \%\ O_2)(N_2\ \text{FORMED}/28)+((CO_2\ \text{FORMED}/44)+(H_2O\_\text{TOTAL}/18)))/((20.95 - \%O_2) \times (1.6028 \times \text{LB OF Moisture/LB OF Fuel})+1))$

2.59

$$\%\ \text{EXCESS AIR} = \frac{\text{LB OF EXCESS AIR/LB OF FUEL}}{\text{AIR REQUIRED}} \times 100$$

= 18.3

TOTAL LB OF H_2O / LB OF FUEL (CORRECTED EXCESS AIR)

= (% EXCESS AIR × LB OF MOISTURE / LB OF FUEL / 100) + LB OF H_2O/LB OF FUEL

= 1.60

(Continued)

Table 4.2 Furnace efficiency calculation

Furnace efficiency with FO firing

Ha	=	AIR SENSIBLE HEAT CORRECTION				
		Cpair × (Tt – Td) × (LB OF AIR/LB OF FUEL)			OR	
		ENTHALPY DIFFERENCE × LB OF AIR OF FUEL				
		C_p air	=		0.24	BTU/LB°F
		Tt	=		80.6	
		Td	=		60	
		LB OF AIR / LB OF FUEL		=	16.94	
Ha	=	83.73	BTU/LB OF FUEL			
Hf	=	FUEL SENSIBLE HEAT CORRECTION				BTU/LB
		Cpfuel × (Tf – Td)				
		C_p fuel	=		0.48	BTU/LB°F
		Tt	=		60	
		Td	=		194	
Hf	=	64.32	BTU/LB OF FUEL			

Hm	=	ATOMIZING STEAM SENSIBLE HEAT CORRECTION				BTU/LB
		Cpmedium × (Tt – Td) × (LB OF STEAM/LB OF FUEL)				
		OR				
		ENTHALPY DIFFERENCE × LB OF STEAM / LB OF FUEL				
		ENTHALPY AT	423	°F	1203	BTU/LB OF FUEL
		ENTHALPY AT	60	°F	1088	BTU/LB OF FUEL
		LB OF STEAM / LB OF FUEL			0.3051	
Hm	=	35.12	BTU/LB OF FUEL			
Qr	=	RADIATION HEAT LOSS				BTU/LB
		63	BTU/LB OF FUEL			
Qs	=	CALCULATED STACK HEAT LOSSES				BTU/LB
		1093.81	BTU/LB OF FUEL			

NET THERMAL EFFICIENCY

$$E = \frac{(LHV + Ha + Hf + Hm) - (Qr + Qs)}{(LHV + Ha + Hf + Hm)} \times 100 = 93.6\%$$

(*Continued*)

Table 4.2 (*Continued*)

FURNACE EFFICIENCY WITH FG FIRING:

Ha	=	AIR SENSIBLE HEAT CORRECTION				
		Cpair × (Tt – Td) × (LB OF AIR/LB OF FUEL)			OR	
		ENTHALPY DIFFERENCE × LB OF AIR/LB OF FUEL				
		C_p air	=		0.24	BTU/LB°F
		Tt	=		80.6	
		Td	=		60	
		LB OF AIR / LB OF FUEL		=	20.88	
Ha	=	103.2	BTU/LB OF FUEL			
Hf	=	FUEL SENSIBLE HEAT CORRECTION				BTU/LB
		Cpfuel × (Tf – Td)				
		C_p fuel	=		0.53	BTU/LB°F
		Tf	=		140	
		Td	=		60	
Hf	=	42.0	BTU/LB OF FUEL			
Hm	=	ATOMIZING STEAM SENSIBLE HEAT CORRECTION				BTU/LB

		Cpmedium × (Tt – Td) × (LB OF STEAM/LB OF FUEL)			
		OR			
		ENTHALPY DIFFERENCE × LB OF STEAM/LB OF FUEL			
		NOT APPLICABLE IN GAS FIRING			
		LB OF STEAM / LB OF FUEL		0	
Hm	=	0	BTU/LB OF FUEL		
Qr	=	RADIATION HEAT LOSS			BTU/LB
		229	BTU/LB OF FUEL		
Qs	=	CALCULATED STACK HEAT LOSSES			BTU/LB
		1816.08	BTU/LB OF FUEL		

NET THERMAL EFFICIENCY

$$E = \frac{(LHV + Ha + Hf + Hm) - (Qr + Qs)}{(LHV + Ha + Hf + Hm)} \times 100$$

		90.0	%

(*Continued*)

Generally, multiple numbers of skin thermocouples are provided; all reading should be normal, i.e., not a single indication should be ignored without knowledge. For example, if one skin temperature reading goes too high and beyond scale, then it can be doubted to be due to thermocouple shoe has given away, of course after due discussion with instrument maintenance colleagues. If one skin temperature indication shows high and within the scale, as a first step, firing in that zone to be adjusted; even if it does not show improvement, then burner in that zone should be extinguished and if required burner in another location to be taken in line. If more burners show higher indication, it should be considered as hard firing in the furnace and overall firing should be reduced to reduce skin temperature irrespective of input reduction is required or not to get the right coil outlet temperature of the main fluid in the process.

4.5 Distillation Column/Absorber/Stripper/Reboiler

We know all the above terminologies in the heading of this section are basically synonymous, for e.g., the section above the flash zone of the distillation column is basically absorption section and zone below flash zone is basically stripping section, and reboiler is also a kind of stripper with one heat exchanger provided at the bottom. Hence, problems in this section are discussed in general.

Now before discussing various issues in distillation columns, let us look back on the principles in the design/operation of the distillation column.

Study in distillation columns was focused mainly on binary components distillation and McCabe Thiele's method is followed and was well demonstrated in undergraduate engineering courses. The method demonstrates mathematically in the case of trays used in the columns. But in petroleum industries, binary component distillation is restricted to only stripper columns like H_2S stripping from hydrocarbon liquid or to the absorber to absorb H_2S from sour gas by liquid amine. Petroleum refineries' main distillation process of crude oil or atmospheric residue deals in multi-component distillation. The method followed here is called the Underwood method and the Fenske equation is used. But solving multi-component distillation design like evaluating a number of theoretical trays in tray distillation column by Underwood method is very rigorous and hence computer programs are used to solve this problem. Next, to evaluate feed tray location, a widely used equation named Kirk Bride correlation is used and the same is also used in computer programmed software.

Now various trays are there in the industries, viz., sieve trays, valve trays, bubble cap trays, specially designed valve trays, etc. Valve trays are most common and economic designs with tray efficiency at about 50 to 60% are reasonable, viz., its efficiency and cost both lies in between sieve trays and bubble cap trays. Next, to get the actual number of trays, a number of theoretical trays are to be divided by the tray efficiency. Following software are available for the design of tray columns:

- Aspen Plus by Aspen Technologies
- Chemcad by Chemstations
- gPROMs by PSE Ltd
- CADSIM Plus by Aurel Systems
- Pro-II simulation

For day-to-day operations, we should understand the following basic parameters in distillation:

- Percent flash vaporization
- Reflux ratio
- Dew point of column overhead vapour
- Percent flooding in the trays
- Steam stripping of column bottom section
  - ➢ Percent flash vaporization is defined as:
    (Amount of vapour generated out of furnace/Total feed flow rate to column) × 100.
  - ➢ Reflux ratio is defined as:
    Ratio of liquid returned to the column to distillate draw rate out of the column.
  - ➢ Dew point of the column overhead theoretically represents the column top temperature considering overhead vapour is not subcooled in the condenser. By solving the equilibrium equation, dew point, i.e., column

top temperature is determined and it can be cross-checked with actual temperature indication.

- Percent flooding at a particular tray, say for e.g., diesel draw tray, is defined as follows:

  (Liquid draw rate of diesel + circulating reflux return (on or above the tray) to that tray + internal reflux) ×100 / (Vapour flow rate to that tray + internal reflux) (4.5a)

  A vapour flow rate is to be determined by subtracting all draw rates below the diesel tray above the flash zone of the column from the total vapour rate entering the flash zone of the column. For approximate calculation of flooding, internal reflux may be ignored for instant guess in day to day operation.

  In tray distillation column, percent flooding generally remains between 60 to 80%, say 75% as optimum. When the value is 100%, it is called total flooding of the tray affecting good mass transfer across that tray resulting in higher pressure drop in the tray and column and higher product draw temperature from that tray resulting in the poor quality of that draw like its FBP would increase.

  When percent flooding decreases below 40%, it is called dumping, i.e., weeping on a larger scale, i.e., it means short-circuiting of tray liquid falling from valve holes instead of flowing through downcomer. This incident results in an unsteady operation like inadequate draw rate and quality from that tray forcing feed rate to reduce.

  As mentioned above that optimum flooding should be 75% can be well utilized to understand what should be reflux ratio in the topmost tray as explained below:

  $L \times 100/(L + D) = 75$

  where L and D are reflux and distillate flow rate.

  From above, it appears that the reflux rate should be three times the flow rate of distillate.

- Steam stripping is provided below the flash zone and that too below the bottom-most tray to strip out the lighter vapour entrapped in the bottom liquid thus helps not only in reducing the amount of residue though by 1 to 2% but also helps in increasing the flashpoint of the residue which is required for safe storage of the residue. The effectiveness of stripping is guided by the partial pressure theory as explained below:

Partial pressure of a component *a*, say,

$$p_a = (m_a / (m_a + m_b + \ldots\ldots + m_n)) \times P \qquad (4.5b)$$

where *a*, *b*, ... *n* represent all components and *m* along with the suffix of the component represent the mole fraction of the respective component, p_a represents the partial pressure of component, *a* and *P* represent total pressure of the column.

Stripping steam constitutes one of the partial pressures; hence, from Eq. 4.5b, it is seen with an increase in steam flow rate, the value of p_a, i.e., the partial pressure of a particular lighter component would come down thus helping in increased vaporization rate of that component. Due to this process working, the bottom stripping is also called partial pressure steam.

Issues in distillation columns operation

In distillation columns, one common problem is once distillate cut draw rate is low and at the same time its density, FBP is also high indicating the draw rate cannot be increased; in some cases, in such situation when delta pressure between flash zone and column top increases, it indicates there is flooding in the draw tray of that distillate cut; even if delta pressure increase is not observed which might be too less to be reflected in measuring scale, the draw rate of that distillate cut should be increased till it is observed that in increasing draw rate tray temperature is decreasing (against the normal phenomenon of tray temperature increase with an increase in draw rate) and the rate increased should be stopped when temperature reversal starts (indicating that flooding of the tray has become normalized); the changes should be done at a very slow rate, say, @ 1 m^3/h at time steps of 15 min to wait to see the effect.

Sometimes, in the crude distillation column, increasing top reflux to reduce top temperature increases top pressure. Hence, there should be an adequate margin in a setting of CDU PSV popping. This may occur when the concentration of non-condensing components like C1, C2 are proportionately too high to show such reflection;

in that case, the top temperature should be reduced either by reducing flash zone temperature or by increasing pump around(s) or respective distillate draw rate(s) in the column.

Sometimes, due to tray dislodgement in any section above flash zone, a short circuit in the absorption section occurs as reflected in lower delta pressure across flash zone and top and poor quality of one cut point and draw rate below the affected tray; similarly, in dislodgment of the tray below the flash zone, residue flashpoint would be low with marginal increase in its yield.

Tray dislodgement problem also occurs in side stripper due to respective stripping steam commissioning in any startup without condensate draining. This results in a lower flashpoint of respective distillate cut even with increasing its stripper steam rate. To survive for time being, input reduction is the only solution or increasing draw temperature if distillation range permits.

Sometimes, bottom stripping steam increased either to compensate for a marginal reduction in furnace coil outlet temperature or to increase flashpoint of residue, but again this should be limited by column overhead temperature and condenser as we know latent heat of water condensation is much more than that of hydrocarbon.

In the vacuum distillation column, sometimes vacuum deteriorates by taking bigger or additional ejectors in line by mistake because additional motive steam limits the condenser thus creating back pressure.

As partial pressure dictates the rate of vaporization, sometimes with an increase of bottom stripping steam, distillates' yield increase even with a marginal drop in vacuum, but this is for fine-tuning and not for bulk adjustment.

Due to specific operating conditions as found in design which causes water condensation in a particular tray in vacuum distillation or another atmospheric fractionating column, a water draw system is provided across that tray; the operating crew should be alert to use that circuit to avoid water shoot up in the column.

In the vacuum column, sometimes distillate draw idle standby pump's suction valve is kept closed to prevent air ingress by virtue of defective mechanical seal or gland of suction valve except for congealing fluid where warm-up circuit is provided to keep fluid circulation in the circuit.

In stripper with reboiler facility, particularly of natural thermosiphon type, reboiler tubes fouling or fouling in the main column, reboiler internals damage even slightly plays a major role in affecting bottom temperature; hence, one should be very careful in overhauling of the column along with reboiler during M&I shutdown.

4.6 Miscellaneous Equipment and Fittings

➢ Valves

For normal isolation and line up purpose, gate valves are used and for the manual control of flow rate globe valve is used where no separate auto control loop is provided. Hence, globe valves are not used much in petroleum refineries, but gate valves are used at random. The problem in a gate valve is in its gate getting stuck up and detached from the stem particularly in water service because of corrosion taking place in presence of water and air; in the hydrocarbon vapour line also the same problem occurs. Only in liquid hydrocarbon service, it does not happen as the joints get well lubricated due to contact of oil with it. To avoid operational problems due to such valve gate getting detached from the steam in closed position, the valve should be laid with its wheel in a vertically downward position in those above-mentioned cases so that gate moves downward by gravity at least if it gets detached from the stem.

- Gate valve should be opened fully to avoid pressure in its stuffing box helping in its increased life without gland leak. Similarly, while closing, it should be hand tight or if the wheel key is used for closing, it should not be over tight closed as in doing so, the valve gate may gets detached from the stem.

Also, the gate valve in gas service should never be kept in a partially open condition as there may be a gland leak from the secondary seal of the stuffing box due to being under constant pressure.

➢ Steam tracing and insulation

Effective steam tracing like covering all zones of fitting installation with adequate insulation is important to avoid congealing of fluid; in this case, selection of steam trap is also very important. Even for non-congealing fluid, right insulation is important to hold its temperature as per process requirement; for e.g., for temperature within 300°C, mineral glass wool type insulation is adequate, but above this temperature, perlite or calcium silicate insulation is required to hold the temperature; however, in industries calcium silicate is preferred than perlite due to lower cost and longer life.

- Instruments

  A common orifice plate is used for measurement of fluid flow. Let us see the principle of working of flow measurement through orifice plate as follows:

  Flow rate through orifice hole is given by:

  $$Q = K \times A_0 \times \sqrt{(\Delta P/(1 - (d/D)^4))} \qquad (4.6a)$$

  where K is orifice constant or orifice coefficient; A_0 is the cross sectional area of orifice hole; ΔP is the pressure drop across the orifice; d is the diameter of orifice hole and D is the diameter of the pipe where orifice installed and fluid flows.

  Here, the orifice coefficient depends on the ratio of d and D, type and proper direction of the bevelled edge around the orifice hole and range of flows and type of fluids used.

  So, the right orifice plate is to be selected to use and the bevelled edge should be in the flow direction otherwise there would be an abrupt wring indication from the flow meter.

  ΔP is measured by the dP cell or other advanced instrument taking pressure impulse from both sides of the orifice. These impulse lines should be as short as possible in length.

  In the instrument setting, generally, the volumetric flow rate, Q which is actually at an operating temperature of flow, is converted to Q at 15°C by density correction; if the actual density of the fluid at 15°C is different from what is provided at instrument setting, then there would be an error in the reading; hence, correction of Q to be done in the instrument setting.

  For measuring gas flow, its impulse tapping should be from the top section of the orifice flange with two tapings upward and at a 45° angle; it should be reversed for liquid flow measurement to get a representative value.

The impulse lines should be well insulated if the sensing fluid is hot. If sensing fluid is not only hot but also congealing, then the impulse lines length should be minimum with effective insulation and should be connected to another non-congealing impulse system through a diaphragm in between to propagate the impulse.

The mass flow meter is costly and used for a specific purpose. Pitot tube type flow measurement is used sometimes for large volumes of liquid, but for large volumes of gas, Venturi-meter gives accurate reading which is costly in CAPEX but without any maintenance.

For level indication, the following formula is used.

$$\Delta P = \rho g L \quad (4.6b)$$

where ΔP is the differential pressure between lower and upper tapping of level impulse lines; ρ is the density of the liquid; g is the gravity constant; L is the level of the liquid.

Here, in the instrument set, density of the liquid is incorporated. If actual density is higher than what is in the setting, then level indication would be higher than what is actual and vice versa thus level indication would be misleading.

Analysers used many times without maintenance do not give representative results due to the common problem of condensation of a vapour or congealing of liquid in an impulse line. Sometimes, calibration also gets upset due to the unfriendly ambience.

- Control valve

  Control valves act as an orifice plate but with a dynamic variation of the orifice hole size depending upon the opening of the control valve. Hence, similar pressure drop correlation like above is applicable here except the value of constant is different and that too varies from types of the control valve selected. The constant here is called control valve coefficient denoted by C_V and the value is provided in all control valve data sheets along with other parameters. The C_V is generally selected in design with maximum pressure drop across the control valve upstream and downstream, i.e., with control valve closed approximately 100%.

- PSV (pressure safety valve):

  A similar concept of flow rate calculation like through orifice plate shown above is applicable here, but in the design of PSV, different formulae are used for a category of fluid flow, for e.g., for gas and vapour flow, orifice area is given by:

$$A = Q \times \sqrt{(T \times Z/M)}/(0.873 \times C \times k_2 \times k_3 \times P) \quad (4.6c)$$

(not applicable for steam).

where Q is the flow rate in kg/h through PSV, T is the temperature of the gas flowing, Z is the coefficient of compressibility of the gas, M is the molecular weight of the gas, k_2 is correction coefficient for constant back pressure, k_3 correction coefficient for variable back pressure, P is the set pressure for PSV popping plus overpressure plus atmospheric pressure, C is the gas constant as given by:

$$C = 520 \times \sqrt{(k \times (2/(k+1)^{(k+1)/(k-1)})} \text{ for } k > 1 \quad (4.6d)$$

$$C = 315 \text{ for } k = 1 \quad (4.6e)$$

where $k = c_p/c_v$, i.e., specific heat ratio of the gas.

In Eq. 4.6c, 0.873 is the valve outflow coefficient which may vary from case to case in design.

As seen above that while selecting a PSV, the following parameter should be considered:

- Fluid to be discharged and its physical state
- Molecular weight
- Ratio of specific heat
- Set pressure
- Overpressure
- Back pressure at discharge, whether constant or variable
- Temperature of the fluid discharged

Hence, once PSV is designed, monitoring that selection criteria are followed in actual operation, else operation parameters should be corrected through analysis.

If two PSVs are provided with isolation valves, the valves should be laid with their wheels vertically downwards as mentioned above. For a single PSV, an isolation valve is not provided but in that case pilot PSV (with its popping set pressure one step lower than in main PSV) is provided with an isolation facility because it is found sometimes that PSV after popping doesn't reset properly which causes continuous flaring of hydrocarbon and prevents to maintain system pressure as desired thus causing unit shutdown on emergency.

For congealing fluid service, generally, a ruptured disc is provided upstream of PSV with a pressure transmitter in between so that rupture disc gives away sensed by pressure transmitter as mentioned above when rupture disc is replaced by isolating the same along with PSV thus avoiding PSV popping; also, in this case, stand by PSV with rupture disc should be provided to take care such replacement job and as isolation facility is provided.

- Fixtures/fittings in the piping

  While fitting fixtures like a control valve, thermowell, thermowell reinforcement pad, quill, bend, support, etc., proper care should be taken. For example, though in the engineering standards norms are provided but in actual execution deviation occurs like, say, there is pipe 90° bending at a very short distance after quill and that too ground support provided at the zone of change in the flow direction, i.e., at the bend; as a result, corrosion took place over a short period of operation. In another example, the bend might be provided at a very short distance from the control valve, which not only caused an error in control valve action or flow reading but also causes corrosion in the bend over a short period of operation.

Chapter 5

Troubleshooting Case Studies in Equipment/Facility Operation

To facilitate the development of knowledge banks in troubleshooting, most of the case studies as experienced have been listed in this book.

To understand systematically, troubleshooting case studies have been categorized into various subsections as discussed earlier and mentioned below:

I. Related to furnace
II. Related to distillation/stripper/reboiler/stabilizer/absorption columns
III. Related to exchangers
IV. Related to pumps
V. Related to compressors

5.1 Furnace

5.1.1 Explosion during Light-Up

- **Incident:** In a furnace, as per procedure, a pilot burner was attempted to light up the furnace, but it could not be lit on. After several attempts and failure, a further attempt was given up and instead the main gas burner was attempted to be taken

Petroleum Refineries: A Troubleshooting Guide
Dhananjoy Ghosh

ISBN 978-981-4877-81-7 (Hardcover), 978-1-003-18040-1 (eBook)
www.jennystanford.com

in line to light up the furnace. In the first attempt, the main burner could not be lit; on the second attempt, there was a heavy explosion in the furnace. Though fortunately there was no human injury, the startup operation was held up and delayed for one week for a thorough inspection and repair, if any. Fortunately, there was no damage to the furnace tube but only explosion doors were to be repositioned.

- **Troubleshooting:** When the burner could not be lit up in the first attempt with the main burner, a good amount of fuel gas remained unburnt inside the furnace box, and thus without proper air purging out of these accumulated fuel gas, lighting up attempted led to an explosion. Also, the attempt to light up the pilot burner failed because excess airflow through the aspirator in the pilot burner extinguished the fire. The art of opening the aspirator is to keep it loosely closed; thus minimum passing air will be adequate to ignite the fuel gas on sparking and also avoid extinguish of fire by preventing excessive airflow.
- **Lesson:** Never light up the main fuel gas burner before lighting the pilot gas burner to light up the furnace.

5.1.2 Frequent Extinguishing and Back Fire during Oil Burner Light-Up

- **Incident:** The oil burner may sometimes extinguish during the lighting up operation. And backfiring also occurs during the same operation.
- **Troubleshooting:** In oil burners, a differential pressure between atomizing steam and oil is to be maintained with the help of a control valve in the atomizing steam header as shown in the diagram as follows:

  In the following flow scheme as available in furnace control, the steam pressure becomes lower than oil pressure by the closing of the atomizing steam control valve due to malfunctioning of differential pressure controller (DPC) which results in backfiring at the bottom of the furnace; so, no person should stand near furnace bottom. Also sometimes, the steam pressure becomes too high due to an abnormal increase in steam control valve opening by malfunction of DPC, which results in the burner getting extinguished. In the present case, when the field controller, DPC was checked, it was observed that the controller was malfunctioning resulting in the hunting of the differential pressure.

- **Lesson:** Differential pressure between oil and steam is to be maintained as per the burner data sheet requirement.

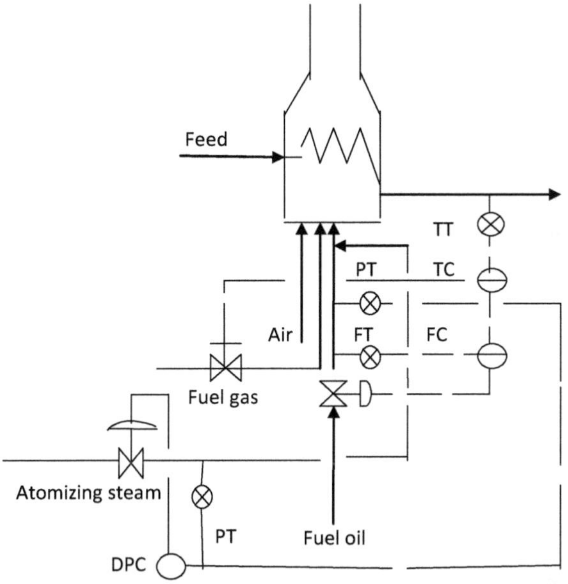

Figure 5.1 FT is flow transmitter, TT is temperature transmitter, TC is temperature controller, PT is pressure transmitter, and DPC is differential pressure controller located in the field.

5.1.3 Furnace Burner Mufflers Getting Damaged Prematurely

- **Incident:** This happened in operation after the M&I shutdown. Flame impingement to muffler observed and two mufflers partially damaged; consequently the flame was touching the floor through the damaged muffler. The burners got cleaned but the problem remained. The operation continued by putting off these two burners, which resulted in lower t'put (feed rate).

- **Troubleshooting:** Switching the burners from oil to gas also could not prevent this flame impingement. The maintenance department was informed about this problem. It was explored whether the refractory of these two mufflers was changed during shutdown or not because if it was done then these two mufflers curing before the start was needed, but it was learnt that there was no replacement in the refractory of the mufflers. Then they were asked to check whether the elevation of the burner tip was as per design or not; while checking, it was found that elevation was lower than design. They could manage to adjust the elevation without shutdown, and thus, the operation could be sustained at higher t'put keeping watch on the burner muffler. In the next shutdown, the mufflers got rectified, cured properly and the problem got solved.
- **Lesson:** This was a result of poor supervision and accountability in the shutdown. In addition, to follow design criteria, there is one tip to check whether burner elevation is correct or not by placing welding rods to the holes in burner which would take the shape of joss stick (*agarbatti*); then observe and ensure by adjusting burner level to see that these welding rods do not touch the muffler.

5.1.4 Oil Burners Getting Coked Up Frequently

- **Incident:** The maintenance crew got fed up and operation steadiness was getting affected. As there was a restriction in increased fuel gas availability, there was no alternative to keep 70% firing with oil burners thus making operation unstable.
- **Troubleshooting:** Flue gas excess air level was checked and found to be alright; furnace draft also was in line. Fuel oil line steam tracing was also in order and working. Fuel oil tank temperature was checked and found normal at 80°C. Fuel oil viscosity was analysed and compared with burner design; it was found that viscosity was about 40% higher than kept at burner specification; the same got rectified by adding diluents in the tank, well circulated before reusing. Meanwhile, a standby fuel oil tank was made ready and taken in line. The problem got solved.

- **Lesson:** If any time there is a deviation in any practice like fuel oil blending ratio and/or temperature, then analysis and approval should be followed before implementing the change.

5.1.5 Furnace Tube Skin Temperature Rising but Oil Outlet Temperature Remaining Very Low during Startup

- **Incident 1:** This happened in the lube hydrotreater furnace; when the coil outlet temperature was not increasing as the operator was increasing the fire, but skin temperature was high; the operator doubted the indication of skin temperature as both the coil outlet temperature indications were low. In this process, the furnace tube got ruptured and the unit got an emergency shutdown with a big fire inside the furnace box.
- **Troubleshooting:** The operator has to understand what should be normal fuel consumption during each phase of start-up. Coil outlet temperature was low as feed to furnace bypass valve was open though upstream feed valve to the furnace was open and thus the only hydrogen was flowing through the furnace tube which is not enough to raise the coil outlet temperature.
- **Lesson:** The HAZOP study of P&IDs was inadequate; the proper P&ID and its execution can prevent this accident in case of such human error.
- **Incident 2:** All pass flow indications and pass outlet temperature indications were normal and out of eight skin temperature indications, only one was showing higher indication. Theoretical calculation of skin temperature was also showing normal value. The decision-making house was puzzled.
- **Troubleshooting:** The burner near the high skin temperature indication was put off and instead, the burner at another location was lit on. In one event, the skin temperature indication got normal and in the other event, it still remained high. In the first event, there was a problem with the burner and it was hitting the skin thermocouple and thus the problem got solved; in the second event, the colour of the tube-outside

surface was observed, which was normal. Thus, the main reason could be that the shoe of the thermocouple might be cracked resulting in higher skin temperature indication which thus was non representative of actual firing condition, hence, in no fear. However, in next shutdown, this skin thermocouple shoe should be rectified.

- **Lesson:** As in the case of the first event above, burner alignment and burner gun should be checked at every opportunity. With regard to the second event, all skin thermocouple inspections inside the furnace should be done during shutdown. If all the tube's skin temperatures are normal except one or two, calculate the theoretical skin temperature or look through a peephole with a face shield. If a yellow or white spot is observed on the tube's skin, then remove the burner from that zone; if still yellow or white spots are seen on the tube's skin then plan for a shutdown to replace the tubes.
  To avoid furnace tube coke up due to low flow, low feed flow security must be inline after establishing feed flow.

5.1.6 Coker Furnace Tube Coking during Total Power and Utility Shutdown

- **Incident:** During total power and utility shutdown, when the coker furnace also tripped, there was a fire inside the box which aggravated during box purging after isolating fuel oil burner valves, thus immediate coil steam purging could not start leading to partial coking up of tubes which however could be cleaned without tube cutting.
- **Troubleshooting:** Through furnace box window peep-hole observation, no spot of oil was found on the tubes. During draining of the coil purging steam header, a lot of black oil was found. This indicates that both check valve and block valve in coil purging steam header was not holding, thus during an unsteady-state situation like power and utility failure, steam header pressure became lower than feed pressure inside coil for some moments leading to backflow of oil in coil steam header; this ingress of oil into coil steam header in turn entered into atomizing steam header leading to the first

fire break out inside box and further aggravated when box purging steam valves were opened.

- **Lesson:** Either coil purging steam header pressure should be higher than feed pressure at coil inlet or special check valve of ensuring zero flow in reverse without fail or there should be double valves with bleed at upstream of the check valve; the bleed should be provided with shutdown valve which should open in auto in such emergency trip.

5.1.7 Furnace Del. P Rising Prematurely after Each Decoking Shutdown of Coker Furnace

- **Incident:** Generally, the coker cycle length of about six months is achievable as per the design and was also being realized; but during the course of time, the cycle length gradually reduced to one month thus increasing energy cost as well as reducing unit availability.
- **Troubleshooting:** It was observed that short residue viscosity, CCR increased due to modification like installation of structured packing in vacuum distillation units (VDUs) which caused a reduction of VR yield from 12% to 10%. To analyse correctly, feed viscosity was maintained the same by increasing VR yields to keep the same coil inlet pressure same at each start of run condition after decoking shutdown; Even then, cycle length could not be improved. It was studied and found that high CCR VR generated due to modification in VDU column when diluted by dropping distillate into it in the column, it was not theoretically possible for homogeneous mixing of distillate with vacuum residue (VR), thus there would always be layering of distillate in the VR tank; as a result, it cannot help in improving cycle length. Hence, it was decided to operate a coker furnace with lower coil outlet temperature even with marginal reduction of tubulizing water injection to the coil, and available unfilled space in the coker drum during coking in there had been partly utilized to increase residence time to enhance coking in there thus to compensate less coking inside the furnace. This resulted in an improvement in cycle length. Hence, during the course of time, the dropping of

distillate to VR stopped, and observed no reduction in cycle length.

- **Lesson:** If feed severity increases with respect to composition, it cannot be nullified just by dilution sometimes, and in that case, operation severity of furnace with such refractory feed is to be reduced, and the opportunity to increase severity elsewhere, if possible, is to be explored, which was the availability of residence time in coker drum here.

5.1.8 High Furnace Wall Temperature

- **Incident:** Box temperature was normal but wall temperature was high; this not only resulted in energy loss but also damaged the paint of the wall and ultimately damaged the panel plate resulting in premature replacement of the furnace panel during shutdown.
- **Troubleshooting:** The operating conditions like box temperature data studied and found that it was always normal; there was no fire inside the box; when explored to investigate the last refractory drying during the previous shutdown, it was found that SOP was not followed religiously. This resulted in premature cracking of refractory with consequent effect as mentioned above.
- **Lesson:** Right refractory and right application along with following design drying procedures are necessary to operate a furnace for the long term. Also, any shortcut should be avoided in operation or in maintenance under any pressure; SOP should be strictly adhered to.

5.1.9 Furnace Tubes with White Spots in Some Locations and Peeling Occurring

- **Incident:** Furnace tubes appearing whitish in colour in peeling location; continuous peeling stopped after rearranging the burners' location, i.e., putting off the burner in peeling location and taking burner in other location. The problem was solved for the time being but the tubes were to be replaced in the next shutdown and burner alignment rectified in the peeling location.

- **Troubleshooting:** The skin thermocouple located at a distant location from the peeling zone could not sense higher skin temperature by flame impingement in the peeling zone; the difference in flame impingement severity by switching from oil firing to gas firing in that location could not be exactly understood. Also, burners got regularly cleaned. The other reason might be the wrong alignment of the burner in the peeling location, which gets rectified as mentioned above.
- **Lesson:** Burner alignment during maintenance shutdown should be religiously followed.

5.1.10 Repeated Fire in APH Combustion Air Duct below Furnace Floor Bottom

- **Incident:** When combustion air temperature was increased to normal temperature after startup, the fire caught up as mentioned above; fire extinguished using a fire extinguisher, oil drained off from air duct and combustion air temperature reduced by 10°C. After some time when the combustion temperature was increased again the fire caught up; the untoward situation tackled similarly as mentioned above.
- **Troubleshooting:** This was happening due to the oil accumulated in the air duct caught into the fire as combustion air temperature reached its autoignition temperature or there was a continuous spillage of oil from the burner into an air duct. Fire at combustion air ducts should not take place if there is no oil in the air duct. Engineering layout design of combustion air outlet to burner should take care to avoid oil spillage to air duct unless drain cap at burner bottom is kept open; but here, the cap was found closed by mistake.
- **Lesson:** All burners bottom caps should be kept open and also the bottom drain valve in the air duct should be cracked open with a connected hose and hose open end to be at a safe location to ensure safe draining of oil (near gas detector location) if by chance oil ingress occurs at the air duct.

5.1.11 Furnace Tubes in Visbreaker/Soaker Unit Getting Replaced Prematurely

- **Incident:** In every maintenance shutdown, 2 to 3 radiation tubes needed to be replaced as per the inspection survey during shutdown due to tubes bowing took place beyond the allowable limit and carburization also observed on lab analysis of the tube cut piece.
- **Troubleshooting:** The radiation tube metallurgy was 5Cr-1/2Mo as per the design and allowable furnace tube skin temperature was 650°C. The furnace tube skin temperature was also close to the allowable limit as mentioned above but did not cross the limit except one or two thermocouples were showing 1 or 2°C higher. But it is also true there cannot be enough number of the thermocouple to take care of flame impingement by the burners; thus flame impingement in some location could not be identified by skin temperature data; moreover, for 5Cr-1/2Mo, the allowable temperature should not be more than 600°C to be on the safer side. Hence, at the next opportunity, tubes were replaced by metallurgy, 9Cr-1Mo, and the problem got solved; also, the practice of the inspection department daily recording flame impingement, if any, was established in the system.
- **Lesson:** There should be adequate cushion in operating limits from design allowable skin temperature to be prescribed by the inspection department, which also should record flame impingement on a daily basis, if any, along with skin temperature and other observations.

5.1.12 Leakage of Economizer Tubes in Visbreaker/Soaker

- **Incident:** In every maintenance shutdown that too at a frequency of two years, the whole bunch of economizer tubes got to be replaced due to severe corrosion.
- **Troubleshooting:** The reasons may be different- one may be due to poor DM water quality, external corrosion by flue gas or tubes, dry running for some time due to water pump failure or delayed starting of water flow after startup. The

internal corrosion due to water quality was ruled out as the same water was circulated in other boilers where there was no abnormality. Flow failure was there but for a very small period and also during the cycle when there was no instance of flow failure the same corrosion of tubes took place. On the other hand, DM water inlet temperature was lower than 110°C whereas flue gas sulphur content was more than 1% wt thus resulting in dew point corrosion and it was observed also that severity of corrosion was at the cold end.

- **Lesson:** To reduce the frequency of replacement, it is better to provide tubes of a higher schedule, but nowadays glass enamelled carbon steel tubes have come into place, and hence tubes at the cold end should be kept as glass enamelled to prevent dew point corrosion.

5.1.13 Optimization Problem with Respect to Excess Air vis-a-vis Furnace Draft in Natural Draft Furnace

- **Incident:** When excess air level increased, stack damper pinched to reduce excess air percentage, but draft deteriorated thus making smoky and unstable flame even after adjusting burner air register.
- **Troubleshooting:** It was observed that there was dual firing in the same burner; but fuel gas burner needs maximum opening of primary air register and closing secondary air register as fuel gas burns easily; for fuel oil firing, it should be reversed as combustibility of fuel oil is less than that of fuel gas. So, burners were segregated and the open-air register of idle burners had been closed first, then stack damper adjusted.
- **Lesson:** Follow segregated firing than dual firing, and first adjust air register, then adjust stack damper. Also, the air register of idle burners should be kept closed to facilitate better draft control.

5.1.14 Yellowish Glow on Furnace Tube

- **Incident:** In the morning t'put was increased to design capacity, but subsequently, it was reported that yellowish light is glowing on one tube. This happened after one burner changed over from gas to oil.

- **Troubleshooting:** The inspection department was called upon; it was doubted whether any pinhole leak generated or not. All skin temperature readings were taken and found to be normal. T'put reduced to 60% and continuous observation on the spot followed. As with pinhole leak, glow should continue to exist and on the other side, with any chance-oil accumulated on the tube, the glow should vanish after some time; hence, the observation was continued. After two to three hours, it was reported that the light glow disappeared. This glow might be during the burner's changeover from gas to oil and oil might have been sprinkled and accumulated on the tube; if there were tar-like material in accumulated fuel, it might burn slowly with a light glow.
- **Lesson:** The fuel oil tank should be prepared by proper mixing of ingredients, and also one should remember that if there is pinhole leak in the tube, then one can go for tube emptying out after shutdown even in hot condition, but with a major leak, emptying can only be done after cooling down the furnace box, and the operation crew should be ready for such eventuality.

5.1.15 Low Feed Flow in Furnace of Furfural Extraction Unit (FEU)

- **Incident:** Feed flow was not increasing in line with proportional control valve opening as per the overall feed flow meter and was affecting unit t'put.
- **Troubleshooting:** It was seen that though the pass flow meter was showing more flow, overall feed flow meter showing less flow; while pass flow meter opening vs. flow reading changes observed, it was found that in one pass, flow showed maximum irrespective of control valve opening; also the pass outlet temperature in that pass was very low as compared to others. The skin temperature indication was out in that pass. Physical observation made through a furnace peephole; it was observed that there was a slight bowing in one tube which

appeared to be in that pass of doubt. The unit was shut down, and it was found that one tube was totally coked up. The tube got replaced and after start-up the unit got normalized; however, in shutdown pass flow impulse lines found choked, got cleaned and rectified.

As per theory, furfural may degrade into coke at high temperatures where iron particles auto catalyse the reaction. Also, for liquid flow indication, impulse line length should be as low as possible to get rid of choking problems which is possible by placing the impulse instrument close to the flow pipe and then filling downstream transmitted impulse lines with non-congealing fluid and extending it to a comfortable location.

- **Lesson:** In furnace, furfural should be prevented from overheating with respect to design as there is every chance of degradation as mentioned above. Also, furfural purity should be strictly maintained and to be without contaminants contributing degradation. Temperature indication should be given more priority for accepting as compared to flow indication.

5.1.16 Asphalt Mix Heater Stack Temperature High in Propane Deasphalting (PDA) Unit

- **Incident:** It was before M&I shutdown of refinery; heater stack temperature was running high at about 300°C; however, it was a natural draft furnace; still it was very high as compared to design of 200°C. Heat recovery in convection section was poor resulting in increase in fuel firing within limit of tube skin temperature. Convection tube was a studded tube originally; it was decided to clean the studded tube in shutdown. During shutdown, it was seen that all studded tubes were severely clogged externally; after external cleaning of one or two studs, it was also observed that there was external corrosion on the tubes. It was decided to replace the tubes, but there was no studded tube in hand, but bare tubes were available. It was decided to replace all studded tubes with new bare tubes, also overhauling of all retractable soot blowers in order to avoid any foul deposit on tube surface during operation. But after shutdown when the unit was restarted, stack temperature was found to further deteriorate to about 320°C.

- **Troubleshooting:** By providing studs on tube, the heat transfer area increases by about 50%; hence, even with clogged studs, contribution of additional surface area by studs was at least partially achievable which vanished due to their replacement by bare tubes; this resulted in further deterioration of furnace stack temperature. Moreover, it was also found that maintenance of soot blowers was very poor and regular soot blowing as per schedule could not be followed. Hence, it was decided to procure studded tubes for the next replacement and thorough debottlenecking of soot blowers in the next shutdown after one year. Also, spalling of refractories to studded tubes surface took place might be due to wetting by soot blowing steam; hence, soot blowing steam temperature vis-à-vis stack temperature was analysed to use steam in superheated condition. The problem was finally solved from the next cycle after performing the tasks as mentioned above.
- **Lesson:** Management of change (MOC) approval system must be followed for implementing any change. Effective soot blowing should be followed, as in many cases improper soot blowing became the major source of energy-inefficient operation.

5.1.17 Furnace Draft Problem in Catalytic Reformer Unit (CRU)

- **Incident:** In a CRU, there are three heaters each one before a reactor in series. All heaters were natural draft fuel gas fired type. At design t'put, draft in the second heater decreased to almost zero and time to time it tended to a positive value which was very unsafe and thus t'put kept lower than design.
- **Troubleshooting:** In the first heater, due to more naphthene content as compared to design, endothermicity in the first reactor was higher which resulted in increased heat load in the second heater. It was also found during the increased load

in the second heater that while burners' tube metallurgy, as indicated by skin temperature, could take that additional load, the furnace box design and flue gas flow dynamics could not cope up with that extra load; consequently, draft deteriorated. The subject was analysed to find some ad hoc solution till the furnace was revamped in due course. Flue gas excess air was analysed and oxygen was found to be 8% which was very high as compared to the normal value of 3 to 4% for natural draft furnaces. It was observed that idle burners air registers were open; those were closed to prevent ingress of excess air. Draft improved directionally; burners' bottom holes were open. For fuel gas burners, the bottom hole need not be kept open, and hence, all burners' bottom drain holes were closed; also, proper sealing of all explosion doors done with asbestos rope; furnace peepholes kept religiously closed. All these actions resulted in negative draft in furnace and the unit could be operated at design t'put.

- **Lesson:** High excess air in flue gas, if any, should be reduced to normal which not only increases energy efficiency but sometimes helps in operating units at desired capacity as above.

  Note: In the CDU/VDU furnace it was observed even with furnace stack damper 90% open from control room, draft could not be reduced. On physical verification of stack damper opening, it was found that as per field transmitter also, opening was 90%; even then the operating crew went up to damper location at top of furnace where it was observed that damper lever got stuck up at 30% opening. The same was rectified with the help of mechanical maintenance followed by instrument maintenance and the problem got solved.

5.1.18 Steam Super Heater Coil Outlet Safety Popping Preventing Flow of Steam to Header

- **Incident:** In the CRU furnace, steam super heater coils provided after convection bank to superheat LP steam to MP steam collected to a vessel outside from where it was led to a MP steam header. While commissioning the steam coils, the vent of the vessel kept open till its temperature and pressure was reached to the conditions of MP steam; then the steam was lined up to MP steam header but it was observed that PSV of the vessel started popping. Thus steam had to be kept venting instead of lining up to the header.

- **Troubleshooting:** First of all when after lining up to header, PSV started popping, it indicates that vessel outlet check valve to header was not holding and if vessel steam could not flow to header, reverse flow of header steam occurred to make PSV popping, and also outlet pressure control valve at upstream of check valve might be malfunctioning. As such, with the heat available in the stack gas, the temperature and pressure was not reaching up to conditions of MP steam; by reducing steam flow to coil, temperature could be reached but pressure was not reaching in vent open condition; vent opening was pinched manually to increase the pressure; it was observed that pressure increased but PSV again popped. The control valve functioning was checked, its sensitivity increased and again venting was stopped and steam led to header. This time after some time again PSV started popping. It was found that steam flow was very less as compared to design flow of control valve and thus control valve could not function smoothly. Hence, steam kept venting till the next shutdown and in the shutdown, the check valve was replaced and control valve also got replaced with lower size. After startup, steam header was commissioned and this time there was no PSV popping.
- **Lesson:** Due to deviation of actual conditions of furnace flue gas from design, the pressure control valve at steam outlet header was not matching with actual flow which basically came down to turn down level of control valve. Moreover, non-holding of check valve at outlet header caused excess flow to reverse direction due to high capacity of MP steam header which caused PSV popping; if check valve did hold, then in case control valve not opening towards header could have opened towards vent thereby avoiding PSV popping.

5.1.19 Furnace Tubes Coked Up in DCU

- **Incident:** During an emergency shutdown, furnace tubes got coked up by inside box firing beyond control due to malfunctioning of various sections. Tube coked detected when coil steam purging was tried but no flow. Spare tube stock was explored but only two tubes were found in the stock. All refineries were asked to find whether they had tubes to spare, but their tubes were not matching in dimension. Procurement was also attempted, but only available through import which was to take a minimum of one month. Taking the unit for a decoking shutdown for a long period was posing a refinery shutdown ultimately.

- **Troubleshooting:** The inspection department was asked to inspect the tubes externally by hammering to find whether it was localized coking or whole tube coking. Internal scaffolding was made to facilitate inspection. As per inspection, localized coking at two/three locations in both passes understood.

  As a result, it appeared that at least four tubes were to be replaced, but two tubes were in stock.

  Decoking by mechanical or hydraulic cutting was thought of. While procurement action was taken, it was decided to arrange for a hydraulic cutter along with a mechanical cutter; none of these two could make it through. Then first some piercing was done through a hydraulic cutter, subsequently mechanical cutter was used to propagate the hole further. In this way, by alternate use of hydraulic cutter and mechanical cutter, ultimately all tubes were made through in 5 days thus saving the refinery from shutdown.
- **Lesson:** Right inventory control should be accountable. Established vendor availability should be explored for such decoking more professionally. Root cause of uncontrolled fire inside the box though identified during the above mentioned situation should be used to take care of such eventuality, if any, in future.

5.1.20 Frequent Furnace Tubes Elbow Replacement in DCU

- **Incident:** In every decoking shutdown, during the inspection it was found tube elbow thinning occurring. This caused not only additional maintenance cost but also reduced unit availability thereby restricting crude t'put in the refinery.

- **Troubleshooting:** It was found that original elbows which were imported lasted long, but subsequent procurement sources were indigenous and these elbows got thinned frequently. Operating conditions were investigated; though coil tubulizing water flow was within design limit, its rate was reduced marginally but no improvement observed during next shutdown. Elbow metallurgy was of P-9 steel; the procured sample piece was lab tested; defect could not be identified. It was decided to import the elbows from the same supplier who originally supplied. With the imported elbows, the problem of thinning got solved.
- **Lesson:** In procurement of critical items, not only specifications of the material but also performance certificate from user should be demanded from the prospective vendor.

5.2 Distillation/Stripper/Stabilizer/Absorption Columns

5.2.1 Lower Distillate Yield and Higher Pressure Drop between Flash Zone and Top in Crude Distillation

- **Incident:** No change in crude characteristics, no reduction of furnace COT, no change in column top pressure was observed, but distillate yields were lower with a consequent rise in residue yield. The top temperature was also normal and so was the top distillate flow rate; Hy naphtha draw rate checked, that was also normal. It was found that both kero and gas oil yields were less, particularly gas oil. Respective circulating reflux flow rates could not be compared with design data as furnace COT was different from design value which was since commissioning to get desired distillate yields.
- **Troubleshooting:** Though the gas oil draw rate was very low, its draw temperature was not low rather slightly on the higher side. Doubting there may be flooding in the gas oil tray, Eq. 4.5a has been applied as follows:

  (Liquid draw rate of diesel + circulating reflux return (on or above the tray) to that tray + internal reflux) × 100/ (Vapour flow rate to that tray + internal reflux)

In the above equation, all data are available except internal reflux data which is to be calculated by mass and heat balance of the diesel tray. However, to get an approximate idea of tray flooding, the internal reflux which is both at numerator and denominator can be ignored.

By applying the formula as above, it was found that tray flooding was more than 100%. Hence, its draw rate was gradually increased; it was observed that its draw temperature was decreasing though draw rate was being increased; during this course of action, after a particular draw rate of gas oil attained, its draw temperature again started increasing with an increase in draw rate which indicated that gas oil tray flooding problem got solved and draw rate was held at that value which was corroborated by applying the formula again. The same phenomenon was not observed while trying with a kero draw rate means there was no flooding in the kero tray. With this experiment, distillate draw rates increased, column delta P and residue yield reduced.

- **Lesson:** Such incidents occur with crude with slop having higher distillate content like diesel cut only.

5.2.2 Crude Distillation Column Top Pressure Increases with Increase in Top Reflux

- **Incident:** To reduce the top temperature, reflux was increased but top pressure increased and PSV popped preventing reduced top temperature.
- **Troubleshooting:** When reflux increased, the mole fraction of C1 and C2 concentration increased being non-condensable thus increasing column pressure and pressure control loop was with condenser and by-pass with split range flow control through the condenser and its bypass and with no release to flare. To explain the situation, Eq. 4.5b has been applied as follows:

$$p_a = (m_a/(m_a + m_b + \dots + m_n)) \times P,$$

where a, b, ... n represent all components and m along with the suffix of the component represent the mole fraction of the respective component, p_a represents the partial pressure of component, a, and P represent total pressure of the column.

In this case, a, b, c, d, e, etc., are actually methane, ethane, propane, butane, and pentane, respectively. With the particular crude processing, it might be that concentration methane and/or ethane which are non-condensable are higher whereas the concentration of propane and/or butane which are actually LPG, and pentane and so on which constitute naphtha are in normal concentration and not higher. Due to this disproportionate composition, the value of m_a and m_b would increase by condensation of $m_{c,}$ m_d and so on thereby increasing the value of p_a and/or p_b whereas values of p_c, p_d and so on would remain the same. Thus total pressure, P would increase as explained below:

$$p_a + p_b + p_c + p_d + \dots\dots\dots p_n = P \tag{5.1}$$

- **Lesson:** Control loop changed to ensure gas flow to column overhead from fuel gas header at low pressure of column overhead and to release to flare at high pressure of column overhead. Then the problem got solved.

5.2.3 Flash Point of Long Residue of Crude Distillation Unit (CDU) Increased to Nearly Storage Temperature of a Fixed Roof Tank

- **Incident:** All operating parameters like bottom stripping steam flow rate, bottom temperature, stripping steam temperature were normal; crude characteristics was also unchanged; however, as different types of crude mix were processed in the refinery, there was chance of mix up, hence, true boiling point (TBP) of crude analysed, found normal. But flash zone pressure was higher.
- **Troubleshooting:** Bottom tray dislodging with condensate carryover with stripping steam during startup was doubted. But it was found that the unit was running normally for a long time after the last startup and in between there was no upset. It was observed that yield of long residue was also higher than normal; while checking against tray flooding as mentioned earlier, a similar problem was found; the flooding rectified by adjusting distillate draw rates in same way as above and this reduced the flash zone pressure and consequently improved flash point of long residue.

- **Lesson:** For stability in column operation, tray flooding should be given first attention.

5.2.4 Flash Point of CDU Diesel Cut Decreased Thereby Restricting Blending of Lighter Distillate with Lower Flash Point to Increase Diesel Production and Threatened Safety in Storage

- **Incident:** Diesel draw rate, draw temperature were normal; crude mix was not changed, no slop contamination in crude, stripping steam rate and temperature in diesel side stripper were normal.
- **Troubleshooting:** Problem could not be solved by changing operating conditions; line contamination was checked, no clue found; this operation continued with low t'put. It was decided to shut down the unit at earliest opportunity and check whether stripping zone trays have been dislodged or not because in unit startup while commissioning stripping steam, if condensate was not drained from stripping line inlet to column, then water being lighter than hydrocarbon would cause huge volume expansion thus creating heavy thrust to trays enough to dislodge the same. In shutdown, doubt proved to be correct which got rectified and the problem solved.
- **Lesson:** Safety aspect in stripping steam commissioning should be followed strictly as explained above and also as found in operating manual of the process unit

5.2.5 Top Vacuum of Vacuum Distillation Column Hunting

- **Incident:** All operating conditions like flash zone pressure, all distillate and circulating reflux draw rates were normal and steady, bottom, top, draw and reflux draw/return temperature were steady. Ejector steam pressure and flow checked but those were also steady.
- **Troubleshooting:** T'put reduced before proceeding to shutdown; but top vacuum hunting did not reduce. Column top zone from reflux up to barometric leg leak survey started and observation in column pressure continued. It was

observed that slope and degree of hunting were deteriorating gradually. When the decision was taken for emergency shutdown, suddenly a hole detected in the barometric leg pipe; immediate quick fixing was done and consequently top vacuum started improving. The unit was brought to shutdown steadily, and the repaired leaky pipe got replaced during shutdown.

- **Lesson:** Survey in unit running condition can continue as long as such hunting is within allowable limit and that too in column top zone only with not a bit deterioration in flash zone vacuum; otherwise, unit should be shut down first and then leak survey should be started.

5.2.6 Dumping of Liquid in Vacuum Column Leading to Unstable Operation

- **Incident:** Vacuum column capacity revamp project was undertaken to increase column capacity by 30% for which internal modifications including replacement of old valve trays with new design was carried out. After startup, it was observed that distillate yields were very low and vacuum residue (VR) yield was very high even at low t'put; as such it appeared to be dumping of liquid which became severe at turn down t'put and did not improve even at increasing t'put. Initially, dislodging of trays doubted which might have happened due to poor workmanship in column trays fixing during shutdown.

  Units shut down, the column opened and found all trays intact and no dislodging took place.

- **Troubleshooting:** Valve tray design calculation checked in site by design authorities with tray flooding calculation with actual operating data following Eq. 4.5a and came to the conclusion that new valves were too lighter to avoid dumping. As a quick fix solution, it was decided to blank off a few trays in each location to be decided by on spot design calculation. The unit restarted after this modification and this time operation found normal but revamped capacity with qualities could not be achieved. It was decided to come up with new design trays in the next shutdown. In the next shutdown, all trays were replaced with new ones and problem got solved.

- **Lesson:** Pilot plant simulation should be done before implementing new trays.

5.2.7 Vacuum Was Not Improving after Startup in Vacuum Distillation Unit (VDU)

- **Incident:** All operating conditions were normal and it was after startup from bottled up shutdown, i.e., the system was not exposed to the atmosphere during shutdown.
- **Troubleshooting:** Overhead condenser cooling water pressure was normal, hence flow also should be normal but no flow meter was there. Cooling water temperature was also normal; ejector steam pressure and temperature were normal, and flow found to be on the higher side. In a vacuum system, three primaries and two secondary ejectors are there. Out of three primary ejectors, one was of smaller capacity and out of two secondary ejectors, one was of smaller capacity. It was found that one bigger capacity secondary ejector and two bigger capacity primary ejectors were in line, because as per design one was kept as standby/spare in each section. Taking spare ejector either in primary or in secondary system in line, vacuum deteriorated further; then design concept of two primary ejectors and one secondary ejector re-established but this time, one big and smaller primary ejector with one smaller secondary ejector taken in line when steam flow came down by about 30% but vacuum improved to normal.
- **Lesson:** When adequate pressure, temperature and flow are required in an ejector system, excess flow of steam creates back pressure due to incomplete condensation in the condensers and thus deteriorates the vacuum. So, the system established since commissioning should be recorded and followed.

5.2.8 Stabilizer Bottom Temperature Decreasing Resulting in Fall of LPG Recovery

- **Incident:** LPG yield on startup after maintenance shutdown was on the lower side. It was observed that column top pressure, temperature and reflux flow/temperature were normal as was before shutdown and the same crude mix processed after shutting down and stabilizer t'put also kept the same. But bottom reboiler temperature was not increasing though reboiler hot stream, viz., gas oil circulating reflux (CR) flow and temperature were normal.

- **Troubleshooting:** Doubting reboiler temperature indication, crude tank with higher LPG potential was lined up, but yield did not improve. It was also found that Reid vapor pressure (RVP) of bottom naphtha products increased which clearly indicated that there was LPG slippage occurring through bottom naphtha. It was also found that the bottom reboiler was cleaned during the shutdown. Functioning of a three-way temperature control valve in the gas oil CR line, i.e., at the hot fluid side of the reboiler checked which was also found normal. Effectiveness of cleaning of the reboiler was doubted as leakage was ruled out due to unchanged distillation property of bottom naphtha found. Hence, unit brought to shutdown and bottom reboiler cleaned and the unit restarted, but the same problem remained. In the workshop, there was a spare tube bundle available with the same carbon steel metallurgy. It was thought there might be a permanent scale developed on running a reboiler during idling of the same in open air during shutdown; the phenomenon has been discussed also in section 4.3. Hence, the unit was brought to shutdown again and the reboiler replaced with a spare one as mentioned above which was also not new. Surprisingly, this time after startup, reboiler temperature increased with consequent improvement in LPG yield.
- **Lesson:** In troubleshooting, no idea should be ignored, and it is learnt that there can be permanent scales developed on heat exchanger tubes due to unknown ambient conditions.

5.2.9 Lower Flash Point of Diesel Ex Diesel Stripper in Diesel Hydro-Desulfurization (DHDS) Unit

- **Incident:** Flash point of diesel reported lower; the exchanger network between diesel stripper and battery limit included only feed vs. product exchanger and finally product cooler; as flash point of feed diesel was normal at about 36°C, whereas flash point of product diesel was lower than 36°C, there was no question of investigating the leakage of the exchanger, if any. Feed inlet temperature and flow to diesel stripper were also normal; no abnormality in stripping steam flow and temperature was observed.

- **Troubleshooting:** Fouling, if any, across the trays measured by pressure survey across top and bottom of the stripper; pressure drop was quite normal. Feed rate to stripper decreased, there was directional improvement in flash point, and more improvement in flash point observed with increase in feed temperature temporarily to check effect as it had to be done by manipulation in the reaction section. Then it was clear that tray dislodgement might have taken place at some point of time, maybe during startup while commissioning steam. For the time being, unit kept running at lower t'put to get the flashpoint manageable; shutdown planned, and during shutdown it was observed that 5 trays severely dislodged in the bottom section; these got rectified and problem solved.
- **Lesson:** Before commissioning stripping steam, condensate draining must be done from a low point drain located near the entry to the stripper.

5.2.10 Heavy Gas Oil (HGO) Ex CDU Column Draw Rate and Temperature Hunting

- **Incident:** CDU column capacity revamp jobs carried out when an additional distillate draw in between flash zone and light gas oil (LGO) implemented to route HGO to hydrocracker unit (HCU) feed tank thereby reducing the yield of residue (RCO). After startup, it was observed that when discharge from the HGO draw pump increased to get the desired draw temperature, after about an hour, the pump started losing suction and flow rate came down and tray temperature increased.
- **Troubleshooting:** Flow control valve function studied; found no problem. The standby pump was tried, but the same problem continued. Design data sheet of the pumps checked and found that both the pumps delivered desired

shut off pressure as well as working pressure at design flow rate. Pumps taken to the workshop and no internal problem found. Pump suction strainer opened; no dirt found. Pressure gauge provided at pump suction; initially, suction pressure normal but after about an hour, suction pressure came down. Then revamp job on HGO zone was explored and analysed; to provide HGO draw facility, no packed bed below HGO draw zone and no chimney tray at HGO draw tray was installed; only a draw nozzle was provided at a particular tray where HGO desired temperature of about 355°C is achievable; the draw nozzle was piped to pump suction directly.

In the next shutdown, the above changes like a packed bed and chimney tray provided and the problem got solved.

- **Lesson:** Follow engineering principles and avoid shortcuts.

5.2.11 Deasphalted Oil (DAO) Flash Point Ex DAO Stripper Reported Low and Reason Could not be Identified

- **Incident:** DAO stripper temperature, stripping steam flow and temperature, all were normal; rundown DAO passed through heat exchanger with DAO-Propane mixture on the other side; to avoid that confusion, hot DAO sample at stripper bottom was collected with the help of a cooling bath; flash point reported low.
- **Troubleshooting:** It was observed that top pressure of stripper was marginally higher, however, there was no scope of pressure going to be very high as top was floated to condensing system without any pressure control; performance of upstream DAO flash separation circuit was checked; no abnormality found; tray dislodgement in stripper was doubted, and hence, feed rate reduced drastically, but no sign of improvement. This performance was after startup from previous maintenance shutdown; hence, all major to minor activities in stripper, all connected lines were investigated; as per record, it was found that all activities monitoring were under control except trifle jobs like steam tracing in the connected lines. It was doubted whether the contract technician had terminated the steam tracing line directly

inside DAO stripper inlet line or not instead of terminating it to a trap; line insulation removed and checking done; doubt was correct; due to this, steam condensate was entering the stripper and was flashing inside. Hence, the unit was brought to shutdown, the line got rectified, the column opened and found two bottom trays dislodged; rectified and the problem got solved.

- **Lesson:** Special care should be taken even for any small-bore job in any shutdown job.

5.2.12 Naphtha Stripper Bottom Temperature in Naphtha Isomerization Unit was Lower

- **Incident:** After idling shutdown for a fortnight, the unit was restarted; all were normal but after some time stripper bottom temperature gradually started falling and holding at lower temperature.
  All operating conditions were normal.
- **Troubleshooting:** It did not improve any more even after increasing steam pressure marginally to bottom reboiler as steam floated to whole refinery thus there was no scope to increase pressure further; decreasing feed rate, however, helped directionally but that was causing production loss. Reboiler steam temperature was normal; its condensate return was routed to condensate header through trap; condensate drained from trap upstream; it was observed that full bore steam was coming out without any condensate. At open atmosphere, of course, condensate flash to steam but also there should have been some condensate along with purge steam while draining in such practice. Some abnormalities inside stripper doubted. As a trial, stripper pressure increased up to allowable pressure with the idea that this higher pressure will help in more flow downwards if there is choking in the trays and if it succeeds then negative effect of higher pressure would be compensated by increasing bottom temperature as there was no problem in the reboiler. This action had improved the stripper bottom temperature and problem got solved. In next shutdown, when stripper was

opened, lots of mucks found in the trays of the stripper; after cleaning, the stripper behaved normal in operation.

- **Lesson:** Choking, if any, in the trays, many times get partly removed by exerting pressure and flow jerk in many occasions, but it was just a chance achievement. However, such muck deposits in the trays are unusual and indicate internal corrosion taking place somewhere or line flushing material in the start-up after turnaround got deposited in the stripper trays. Hence, flushing in start-up should be strengthened.

5.2.13 Higher H_2S in Light Naphtha Ex De-Ethanizer Column in Hydro-Cracking Unit (HCU)

- **Incident:** The unit went for capacity revamp shutdown; after restarting the unit, H_2S content of light naphtha could not be brought down thus making it unsuitable for blending in MS pool, for processing as gas turbine (GT) fuel in power plant, and also making it unsellable as naphtha product. The column operating conditions were normal. If its bottom temperature was raised by increasing reboiler outlet temperature or by decreasing pressure of the column, then naphtha H_2S content got reduced but weathering property of LPG went high and also H_2S content of LPG went high.
- **Troubleshooting:** It was observed that there was no scope to adjust operating parameters in de-ethanizer, and also tray dislodging of the column could not be doubted as reversible phenomena as mentioned above occurred. Feed sulphur content was explored and found to be normal. Upstream flash separators in the reaction section also found to be normal. However, in the main fractionators, operating conditions in top zone, viz., kero draw rate, top temperature, top pressure, and overhead offgas compressor operation were not steady. In kero stripper, levels were getting filled up and emptied out in cycle and for that its level controller had been found to be in no fault rather its inlet flow from main fractionator was unsteady. It was seen that to arrest excess H_2S ingress to de-ethanizer, top temperature of main fractionators had to be kept lower by

enhanced reflux flow and in that overhead offgas compressor was starving in suction due to less flow making flaring of offgas and losing LPG yield; so, top temperature of main fractionator increased to optimize suction flow to offgas compressor by intermittently adjusting top reflux flow to main fractionator, but this time top temperature kept as minimum as possible so that resultant flow to de-ethanizer can be kept as low as possible and offgas compressor can be run steadily; but all these done with continuous manual intervention in control panel. With this, with marginal increase in reboiler temperature of de-ethanizer, H_2S content of light naphtha could be controlled within limit simultaneously keeping LPG yield and quality also within tolerance. Operation was sustained in this way till the next shutdown, and it was concluded that there was some error in engineering inside main fractionator during revamp shutdown. During the next shutdown, the kero draw trays had been modified and its layout with respect to withdrawal pan also got rectified. This resulted in a steady liquid level in the kero tray resulting in steady kero draw rate ex fractionator which ensured steady exit of H_2S along with kero flow thereby reducing H_2S load in overhead keeping desired flow to offgas compressor. The H_2S content of light naphtha also got solved.

- **Lesson:** Many times, re-engineering in a fractionating column found to cause problems in the next operation. Hence, operating engineers should be engaged with the designer at the engineering stage if pilot plant distillation cannot be simulated.

5.2.14 Intermittent Fire from Column Shell of Hydrocracker Main Fractionators

- **Incident:** After starting the unit from revamp shutdown, a smoke from column shell observed; extinguished through fire water hose; again after a few days same thing observed; it was presumed there might be some oil soaked in the insulation, so exercise repeated to douse the smoke. After a few days, not only smoke but small fire was also observed; this time after extinguishing the fire, insulation was removed from that zone.

But again after a few days, fire was observed from the same location where there was no insulation.

- **Troubleshooting:** As fire took place from the zone without insulation, nearby insulation was checked and found to be dry, and no hole observed in that zone; it was puzzling; in the running unit, inspection from close proximity could not be carried out. Hence, it could be concluded that there might be some pin hole in that zone where fire took place intermittently. As per inspection thickness survey report during shutdown, there was no deterioration in thickness in any survey zone and also no internal corrosion spot was observed in the fractionator. Hence, it might be due to some lapse on the part of the welder/fabricator during shutdown who worked inside in that vicinity during the previous shutdown. Considering this, to continue operation with safety and as well to enable to take a planned shutdown, it was decided to place a steam lancer in that zone; this stopped repetition of fire. In the next early shutdown, a thickness survey was done from outside in that zone and found the thickness was okay and it was a pinhole there. To avoid a long time shutdown for correcting from inside, it was thought that pinning technology application from outside in that zone should be adequate to stop the leak and with its success, thorough repair from inside can be done in the next turnaround. The idea worked and in the next operation fire did not take place even without steam lancing; also insulation was put in after pinning.
- **Lesson:** Test sparking of welding rod might have been done on the column shell by the welder during shutdown, and always they should be educated to use test kits for this purpose. Also, the inspection department should thoroughly inspect the column shell again after any job done, if any, inside the column.

5.2.15 Level in LPG Gas Absorber Rising in HCU Even with Spare Bottom Pump Running

- **Incident:** Again such abnormal operation observed after shutdown. Normal flow of light naphtha was maintained to the absorber column for H_2S absorption from LPG in the absorber. There was one bottom pump running with one spare to maintain the level. The pump was changed over to spare one but the same problem remained. There was minimum circulation line from pump discharge to suction without isolation valve but with a restriction orifice in the circulation line; as pump was of very small capacity of 9 m^3/h, no special care like circulation control valve was provided in design and circulation return line was connected to pump suction at downstream of suction valve; no suction strainer choking found; no maintenance was done to the pumps. The operation could be sustained with both pumps running. But when one day one pump was giving problems and was required to be sent to the workshop, it became difficult to sustain operation. The pump with problems like mechanical seal leak given to maintenance and t'put was reduced to sustain operation.

- **Troubleshooting:** It was found that the column bottom line up to pump suction was opened during shutdown to flush the column; so labours were engaged in that area. Also, the pump did not starve in suction, rather a single pump was tripping in operation; also, no problem in the motor found. So, there was no question on choking of pump suction and circulation line, rather flow load was higher. Feed rate checked; found to be normal; it was also found that the restriction orifice in the circulation line was in position, but there was hole in the tail of restriction orifice. The same opened and found that it was not like a restriction orifice rather like a metal gasket. It was doubted that the technician during shutdown while cleaning the line must have lost the restriction orifice and managed to box up the line with a metal gasket with a tail without hole in the tail but with inner hole diameter equal to pipe ID (internal diameter) without informing anybody. The pump datasheet was studied and the desired orifice put into the place, and the problem got solved.
- **Lesson:** In a shutdown, contract workers should be educated not to conceal any mistake made by them, and they should remember that confessing a mistake is better than hiding it from safety and operational point of view.

5.2.16 In Naphtha Hydrotreater (NHT) Unit, Column Circulation was Taking Time Delaying Feed Cut in

- **Incident:** In pre-commissioning, thorough cleaning of the column in the recovery section was done as per the manual. Then cold circulation was started with 200 mesh strainers in pump suction against a normal mesh size of 40 in normal operation in order to save the pump from damage debris coming out, if any, from the column during start-up circulation. After it was observed that there is no fouling in the strainer, hot circulation in the column circuit started while simultaneously reaction temperature was being increased in the reaction section in order to make it ready for feed cut in. But during this period, it was observed that the pump suction strainer in the column bottom was getting frequently choked and being cleaned frequently. Thus feed cutting was getting delayed as the recovery circuit was not getting ready.
- **Troubleshooting:** The scale found in the pump suction strainer during hot circulation was very fine; this was shown to maintenance people to confirm whether this can damage pump internals or not. They agreed to replace this 200 mesh strainer with a strainer size of 80 mesh unless there is no problem with the process point of view. This start-up naphtha was not routed to the feed tank and instead it was routed to product naphtha tank; hence, there was no question of feeding dirt naphtha to the reaction section. So, a strainer of 80 mesh was put in and an uninterrupted operation of column circuit established. Thus the problem got solved. Naphtha being a cleaning agent will always scale-out from column shells as it is more and more heated.
- **Lesson:** Fine mesh for cleaning should be provided only during cold circulation.

5.2.17 Asphalt Stripper High Level Problem in Propane De-Asphalting (PDA) Unit

- **Incident:** May a time, high level problem in the stripper occurred due to foaming in asphalt; operating crew were habituated like increasing bottom pump stroke rate by increasing motive steam flow to the turbine of this positive displacement pump, correcting the antifoam dosing rate or increasing antifoam solution strength or changing over to standby antifoam pump. This did not solve the problem and resulted in lowering the t'put thereby not only decreasing bitumen production in the refinery but also creating ullage problems in the feed tank, and consequently threatening refinery crude t'put. This happened in operation after the M&I shutdown.

- **Troubleshooting:** Sometimes when increasing the stroke rate of the bottom pump did not help, the standby bottom pump also was taken line; when both antifoam pumps failed, the problem in the pump identified, rectified and the problem used to get solved. This time, all such actions failed to yield any result. A doubt occurred whether the upstream propane flash separator was functioning well or not; but it was observed that pressure, temperature, load in the separator was normal; also, flashpoint of rundown asphalt was found normal, so the question of upstream flash separator malfunctioning got eliminated. Level control valve functioning was also checked and found to be okay; the maintenance department was asked to check functioning of the bottom pumps, but in vain. Choking, if any, in downstream line was also thought of, but such eventuality was ruled out after one/two days as there was partial flow always, so, partial choking, if any, should have been cleared by that time because rundown asphalt temperature was high at about 200°C and line was steam traced. This line was routed to the bitumen unit; hence, line survey was carried; it was found that the asphalt line from stripper bottom pump discharge was routed straight to the bitumen column going up to its top and from there it came down to bitumen bottom pump bypassing the column as column was necessary to operate only during air blowing operation when vacuum residue (VR) was taken instead of asphalt to produce bitumen. Bitumen bottom pumps also were functioning normally in series; bitumen blending components like aromatic extract were mixed on line, then combined line was routed to rundown through a steam generator. This line up was checked, all valves were found open, but still flow restriction. It was found that the antifoam solution expiry date is over, hence, request given to the commercial department to purchase fresh chemical.

Suddenly, the question came to go up to the bitumen column top as the line was routed through there but the troubleshooting group debated that if there was any valve at top and it was in closed position there should not have been any flow through the line. Keeping this logic in abeyance, still it was decided to go up to top and check; it was found that valve at top was in close position, and the flow was occurring due to passing of the valve; hence, the valve got opened fully. The valve was closed and blinded during shutdown, the blind was removed before startup, but the valve was not opened. The new antifoam solution which was ordered reached the site by then but it was no longer required.

- **Lesson:** All remote acceptability points should be taken into consideration in troubleshooting, and accountability in following start-up SOP should be in place.

5.2.18 Vacuum Stripper Level High in Lube Hydrofinishing Unit

- **Incident:** Stripper level went high after startup from maintenance shutdown; all operating parameters were normal. T'put reduced which restricted production of Lube and also caused feed tank ullage problem, and threatened refinery crude t'put if continued for long.
- **Troubleshooting:** Liquid viscosity checked, found normal, bottom pump functioning was normal; there was a backflow control valve; though it was functioning alright, its isolation valve was temporarily closed to check level depleting or not, but in vain. The bottom of the vacuum stripper was routed back to the steam stripper located below the vacuum stripper; there was no level problem in the steam stripper. The bottom line of the vacuum stripper was hot, so the line was not choked and partial flow was there. However partial clogging of that line by debris in shutdown was raised, but the operating crew did all bottom line flushing full bore as a pre-startup and pre-column box up activity; hence, that question was ruled out.

The question of tray dislodging in vacuum strippers by steam-condensate hammering was also ruled out as there was no stripping steam provision in the vacuum stripper; however, poor workmanship in tray boxing during shutdown was debated but the supervising officer ruled out this. Finding no clue, it was decided to go for shutdown again and open the column for checking. After opening the column, it was found that the vortex breaker at the column bottom got detached and laid flat partially closing the bottom pipe hole restricting the flow. After rectification of this, the problem got solved.

- **Lesson:** There was a poor level of inspection during the shutdown. SOP for inspection activities should include checking of vortex breaker also along with conventional practices.

5.2.19 Unsteady Top Temperature of Solvent Drying Column in Furfural Extraction Unit (FEU)

- **Incident:** In solvent drying column, all dry furfural hot streams from two flash separators routed to solvent drying column where wet furfural-water mix was used as top reflux; the wet furfural vapour from its top joins wet vapour from top of furfural-water azeotrope distillation column and finally routed to a vessel through overhead condenser for phase separation into two phases, viz., furfural rich wet solvent and water rich azeotropic wet solvent; this furfural rich wet solvent was used as top reflux to solvent drying column as mentioned above and water rich azeotropic solvent routed to azeotropic distillation column as feed to it; finally, the bottom of azeotropic distillation column routed to open drain which is free of furfural smell.

  Once during operation it was observed that pressure in one flash column started increasing to a value after which it automatically reduced to original value and this was repeating in cycles, and there was furfural slippage from azeotropic distillation column bottom.

- **Troubleshooting:** Due to above, solvent drying column top temperature was fluctuating as top reflux could not cope with such cyclical change; this in turn was affecting temperature of azeotropic separator vessel as mentioned above and thus changing composition of azeotropic distillation column feed with final consequence of furfural slippage from azeotropic distillation column bottom. The pressure fluctuation was in the high pressure flash vessel which was post furnace after heating it to about 225°C for effective flash; but this furnace was prone to degradation of furfural to coke at that temperature even without presence of oxygen if furfural quality is not up to the mark beyond detection criteria as per quality specification; also, once furfural coke generated in normal ambience due to presence of oxygen, it catalyses further degradation at high temperature even without oxygen in presence of iron particle. This coke partially clogged the flash separator overhead line near an uprising elbow which was causing the cyclical change in flash separator pressure and drying column top temperature resulting in furfural loss plus water pollution as mentioned above. The cyclical change in pressure can be well understood from the flow sketch as shown below:

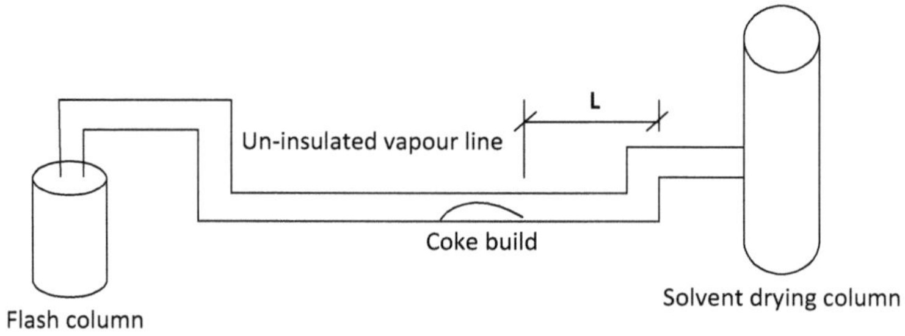

Figure 5.2 Effect of piping layout in process unsteadiness

In the above flow scheme, pressure of the flash column is 0.5 bar and that of the drying column is 0.1 bar, which means differential pressure is very low at 0.4 bar. Due to the coke build up in the location shown above and that line is not insulated, during vapour flow partial condensation of vapour would occur with gradual increase over time and a liquid pool would get locked up in the zone, 'L' as shown above thereby reducing the area of flow of vapour resulting in pressure increase of the flash column; after pressure increase by 0.16 bar of flash column as observed, the pressure suddenly got reduced to its original pressure of 0.5 bar which means with additional pressure of 0.16 bar, the line vapour could blow out the liquid pool to the drying column.

It was thought how to survive operation for the time being and take up the pipe cleaning job or pipe layout geometry change to fully downward instead of existing u-bend path. To continue operation, a software alarm was given to DCS to alert the operator before high pressure breaking in the flash separator; this helped the operator to increase drying column reflux in manual mode a bit advanced thus avoiding fluctuation in top temperature of drying column. Finally, furfural slippage could be controlled this way and shutdown planned to clear the clog as above. Post shutdown operation was free of any problem.

- **Lesson:** Change in furfural specification was brought in like the past supply source was looked into when there was no such problem. The feed source of making furfural during these two periods were explored by talking to suppliers, and next time onwards in procurement, feed source of making furfural was incorporated, and the manufacturing process of double distillation was given thrust in procurement.

5.2.20 Amine Make-Up Post Regeneration Increasing in Amine Regeneration Unit (ARU):

- **Incident:** Sour water examines absorbers of different units entering the amine regeneration column. It was seen H_2S content in regenerated lean amine was higher than the desired limit and also the strength of amine was decreasing thus resulting in a higher makeup of fresh amine. The operation was manageable by increasing column bottom temperature to reduce H_2S content but amine consumption/makeup increased.
- **Troubleshooting:** There was one mechanical filter and one activated carbon filter to clean the regenerated amine to avoid its degradation thus controlling fresh amine make up. As per design, total regenerated amine was flowing through

mechanical filters and then 10% of total flow was passed through activated carbon filters. Cleaning frequency of mechanical filter was as per schedule and no such abnormal dirt observed during cleaning. It was decided to increase the flow through activated carbon to 20% of total against 10% normal. Amine make up directionally improved but not up to the mark. Also, this action led to earlier deactivation of activated carbon. Hence, the operating crew could not be happy. Meanwhile, it was also observed that there was oil carryover to rich amine fed to regenerator. The upstream units were investigated and finally the cause was eliminated. This ultimately helped in avoiding amine strength reduction post regeneration. Higher feed rate to activate carbon was maintained throughout the cycle to quickly purify the amine. In the next cycle, after replacement of activated carbon, flow through that bed was normalized to 10% of total flow. Also, in the next cycle close monitoring on activated carbon was carried out to evaluate its efficacy as it is known that the efficiency of activated carbon depends on the source from which it is made. Best experience of using activated carbon from a coconut shell was used to compare.

- **Lesson:** Sometimes there may be dual reasons of failure and hence, in troubleshooting exact cause does not become identifiable. So, there is nothing wrong in taking two fold actions till the cause zeroed down.

5.2.21 H_2S in Stripper Bottom Outlet Increased in Naphtha Hydrotreating Unit (NHTU)

- **Incident:** This happened on start-up after an idle shutdown of the unit. There was a sulphur guard in the downstream which used to be in line to avoid sulphur slippage to downstream CRU. But after start-up stabilization of the unit, this sulphur guard was supposed to be bypassed to reduce frequency of sulphur guard bed replacement; this was only to take care of start-up sulphur shock. But this time it could not be bypassed due to sulphur slippage continuing.

- **Troubleshooting:** Stripper bottom temperature was normal; this was increased to strip off excess H_2S but in this hydrocarbon loss to fuel gas increased through the top pressure control valve. Similar situation arose when stripper pressure was reduced instead of temperature. Upstream reaction section was investigated and found to be normal. Feedstock crude source was unchanged, still feed tank sample analysed to find the level of total sulphur and H_2S. It was found that total sulphur was high and there was presence of H_2S in the feed tank Naphtha; elemental sulphur content level was normal. Upstream unit operations were explored; naphtha from CDU and HCU are mixed in this feed tank. While there was no question of H_2S slippage to straight run naphtha, it was observed that HCU light naphtha which was routed to this tank contained free H_2S as a de-ethanizer in HCU was not working properly. It was really troublesome to solve the problem of de-ethanizer which started after revamp in HCU.

  Hence, temporarily the pressure of the flash separator in naphtha hydrotreater unit was marginally reduced keeping watch on recycle compressor suction pressure and consequent rise in its discharge temperature; there was enough margin in allowable discharge temperature of compressor. Thus the problem got solved for the time being and got fully solved after HCU shutdown and rectification of the problem in there.
- **Lesson:** Sometimes when there is limitation to eliminate root cause problems, some ad hoc solution also should be looked into if operation sustainability is possible.

5.2.22 High Gas Oil Sulphur Level in CDU Restricting E-III HSD Production

- **Incident:** Before 2017, there were two grades of diesel to be produced by refineries, viz., E-II grade for the rural areas, and E-III grade for the city areas. Refinery was designed to produce E-II grade only. When demand came from marketing to supply 300 to 400 TMT per month of E-III diesel, modification investment became a natural option, otherwise refinery t'put to be reduced as a part of straight run diesel was to be processed in HCU thereby consuming its capacity and ultimately leading to lower crude t'put.

- **Troubleshooting:** As per design, CDU sulphur content was 1600 ppm and actually about 1500 ppm was achievable. So, blending of this straight run gas oil to diesel pool was limited; blending of kerosene was also tried but restricting cetane number of diesels. It was observed that as per design, kerosene draw and circulating reflux rate was higher than gas oil draw and circulating reflux rate. So, it was thought though the cetane index of kerosene was low but its tail end should not be that low. Hence, operating conditions of the column were modulated like reducing both kero draw rate and its circulating reflux; simultaneously increasing diesel draw rate and its circulating reflux. It was seen that there was no flooding in the trays and the column was stable. In this process, sulphur content of gas oil reduced to the level of 1000 to 1100 ppm max and also cetane index of gas oil only marginally came down from 41 to 40. This helped in more blending of CDU gas oil to diesel pools thereby facilitating production of E-III diesel as per demand without reducing crude t'put and without any investment. As cetane Index became normal, the cetane number problem in any case could be tackled by dosing cetane improver as allowed for refineries in northeast India.
- **Lesson:** Sometimes quick win solutions to problems avoids investment and time; however, as seen from above, sometimes achievement is limited in certain cases.

5.3 Heat Exchangers

5.3.1 Product Run Down Cooler's Leak Could Not Be Identified and Cooling Tower Oil Content Became Alarming

- **Incidents:** There were a minimum of 10 process units and as many as 100 products run down coolers for cooling respective products, and the cooling tower was common both for process units and power plant. One day it was observed that there was oil ingress into the cooling water sump. Initially, it was tolerable without being able to detect leak sources. After 2 to 3 days, the oil content in the cooling water sump increased and obviously, its cooling efficiency decreased. As a result, process units' products run down temperature increased, but it could be tolerated as product temperature was lower than flash points of the respective products and also were lower than 100°C for products storage with fixed roof tanks; but lower cooling efficiency of this oil contaminated water was affecting seriously in power plant where cooler was routed to vacuum condenser connected to back pressure turbine of the steam generator thereby affecting power generation capacity. While labours were engaged to skim out the oil from the sump, simultaneously all process units coolers were checked against the leakage of oil to cooling water. But the skimming process was neither successful nor any cooler leak identified. Hence, this was gradually leading to a situation of power tripping and leading to refinery shutdown.

- **Troubleshooting:** After collecting skimmed oil sample, the same was sent for laboratory analysis; initially oil properties were similar to those of lighter oil like diesel, but after a few days when leak aggravated and adequate skimmed oil sample could be collected, the analysed properties indicating leakage in cooler(s) with high viscosity oil; this result in turn indicated leakage in lube oil coolers. As it was a lube refinery, there were a lot of lube oil coolers; all checked from respective bleeders in cooling water outlet lines, but no appreciable oil found in the water sample; the oil content in cooler outlet water was more or less the same as in cooler inlet water sample. When it was totally puzzling, it was decided to attack the most doubtful cooler in line with the properties obtained from analysis of cooling water sump samples.

  An aggressive field test was carried out with most doubtful 2 to 3 coolers like gradually closing the cooling water outlet valve of the coolers and water outlet bleeder points kept open; with this process, risk was oil temperature to be going very high and while reopening the cooling water valve there may be hammering in water line with possibility of steam entering outlet water, but as the experiment was needed for very short period, risk was decided to be acceptable. With this experiment, within a minute there was huge oil coming out in the cooler outlet bleeder from a particular product cooler. That particular cooler was isolated, and the respective unit had to be shut down as it became difficult to operate the unit without that cooler.

This ensured prevention of other units and the total refinery from shutdown.

- **Lesson:** It is safer to have a separate cooling tower for power plants. Also, cooling water (CW) bleeder tap should reach up to pipe central position as with smaller leak, the oil may move through pipe centre line due to fluid inertia point of view.

5.3.2 Frequent Choking in Shell Side (Furfural-Raffinate Mix) of Exchanger in FEU

- **Incident:** The exchanger was to exchange heat from extraction raffinate product in the tube side to raffinate-furfural mix in the shell side before sending the raffinate mix to the furnace. With the progress of time, it was observed that heat recovery in the exchanger was getting poor which resulted not only in increased furnace load but also increasing load in raffinate run down cooler. Sometimes it happened such that shell side had to be bypassed to maintain flow of raffinate mix which was worsening the situation as mentioned above. In every shutdown, when the exchanger was opened it was found that the shell side was severely choked. This became a perennial operational problem in the unit.
- **Troubleshooting:** Exchanger design was checked with respect to operation just after M&I shutdown; it was found it was absolutely okay during start-up operation and matching with design, but gradually it was choking. There was a furfural degradation problem in the unit forming coke-like fouling product, and this issue could not easily be solved; the raffinate product rundown sample was a clean hydrocarbon unlike raffinate-furfural mix. It was explored to see whether

upstream raffinate-furfural mix pump can take care with marginal higher delta pressure in tube side as compared to that in shell side or not; it was found that pump side there was no problem; then heat exchanger calculation done interchanging the flow position of raffinate run down vs. Raffinate-furfural mix applying the correlations described in section 4.3. It was found that the changeover would be perfectly okay. Hence, it was decided to interchange the flow path to ensure more velocity in the raffinate-furfural mix side to avoid coke deposition. After implementing this change in the next shutdown, the problem got totally solved.

- **Lesson:** The fluid which increases fouling during the course of time shall be preferably in the tube side of the exchanger if pressure heads permit.

5.3.3 In Steam Heater, Temperature Not Rising Affecting Solvent Recovery in Solvent Dewaxing Unit (SDU)

- **Incident:** In SDU, dewaxed oil-solvent mixture passes through a series of heat exchangers before entering flash separator for solvent recovery. In that series, there was a steam heater where oil-solvent mixture passes through shell side and medium pressure steam passes through tube side. The temperature of the oil-solvent mixture at the outlet of the heater was not rising thus affecting solvent recovery. T'put had to be reduced as well as a solvent ratio reduced sacrificing oil yield to sustain the operation and to avoid shutdown. This resulted in feed tank ullage problem and posing threat to maintaining refinery crude t'put.
- **Troubleshooting:**

  Case I: There was no individual steam flow meter for that steam heater instead there was an overall steam flow meter in the unit. It was seen overall steam flow was lower than normal at full t'put. Trap outlet condensate bleeder was open; mixture of steam-condensate came out but difficult to understand adequacy of steam-condensate flow. Leak was doubted in the heater, but in that case, in solvent flash column pressure surge should have occurred due to steam ingress to the oil-solvent side which did not happen at all. There was a bucket type steam trap at outlet of heater; the trap was isolated and bypassed to maintain the steam flow; but in this process, there was marginal rise in temperature because steam was not getting adequate back pressure to condense, moreover hammering in condensate line started; so, this operation could not be continued. Then, it was decided to open the bucket trap to check whether there was a problem inside or not. Two long hoses have been connected at two bleeders upstream of bucket trap and bleeder valve sufficiently opened for open draining of outlet steam-condensate and bucket trap isolated and opened; it was found that bucket inside was in inverted position; the same rectified and problem got solved.

Case II: In another steam heater in the same unit, it was found that the trap was in the wrong direction; the problem got solved after rectification.

Case III: In the third case in other units, it was experienced that poor heating in the heat exchanger was due to malfunctioning of the trap which got resolved after trap replacement.

- **Lesson:** Problems in traps are very common phenomena and all sorts of checking should be done as mentioned above.

5.3.4 In DHDS Unit, Heat Exchanger Outlet up to Furnace Inlet Line Was Vibrating after Commissioning Unit

- **Incident:** The heat exchanger service was to exchange heat between gas oil-hydrogen mixture in shell side and reactor effluent mixture in tube side. The shell side outlet was routed to the furnace inlet as usual. Initially, there was no vibration; after increasing load, vibration started and created doubt on continuation of operation.
- **Troubleshooting:** Furnace box checked; firing was normal and all conditions on furnace operation also were normal. Flow meter of circulating hydrogen was checked whether it was in excess or not; with limited possible checking of the flow meter and observing other parameters of the recycle compressor, there was no doubt on normalcy of hydrogen flow. Pipe supports were checked through and through; no

abnormality found. Engineering department asked to check adequacy of piping support in design; no problem found. It was referred to engineering procurement and construction management (EPCM) to check on their side. EPCM contractor reported they did not find a problem in piping support design and execution, but they wanted to open the heat exchanger as they got some doubt in design, viz., design/layout of baffle at shell side inlet might cause sinusoidal vibration effect in the downstream. Unit brought to shutdown, exchanger opened, shell side inlet baffle layout modified as well as modification carried out in other baffles in the shell side. After rectification, the unit started again and there was no problem.

- **Lesson:** HAZOP study generally does not include checking this aspect; hence, there should be a design stability check list in HAZOP study.

5.3.5 Vacuum Residue (VR) vs. Atmospheric Residue (RCO) Exchanger in VDU Choking in VR Side

- **Incident:** VR was on the tube side and RCO was in the shell side. During the course of time since start of operation after M&I shutdown, heat exchange deteriorated resulting not only to increase load in the furnace but also increased VR run down temperature affecting other unit operations using VR as feed. This ultimately resulted in a decrease of t'put in VDU which in turn reduced refinery crude t'put.
- **Troubleshooting:** During shutdown, it was observed that the tube side was partially choked but tubes could be cleaned and hardly 5% tubes were required to be plugged. It was also seen that in every cycle of operation after M&I this was happening; cleaning in shutdown was also monitored, but problems remained. The design vs. actual velocity in tube side checked, no abnormality found. Finding no clue, it was decided to do minimum modification in the tube side to increase velocity of VR during shutdown without going for long time procurement action for a new modified exchanger. The best solution on this line appeared to be to place tube inserts in the tube side, and the agency was lined up accordingly. This modification

contributed to more turbulence in the tube side thereby preventing fouling and ultimately solved the problem.

- **Lesson:** For heavy viscous fluid in tube side, tube inserts are the easiest solution, but use of helical exchanger, which however requires total replacement of old exchanger with new one, found to give better long-term solution.

5.3.6 Gas Oil Cooler in Visbreaker Unit (VBU) Leaking Frequently

- **Incident:** Hot gas oil coming out from fractionators at about 270°C after cooling to about 50°C in a cooler before it is routed to absorber to absorb H_2S from sour gas and before sending it to the DHDS unit. The cooler tubes leaked frequently, viz., once/twice a year; the problem tackled by plugging the leaky tube(s) but in this process when the number of plugged tubes reached more than 10%, Inspection department recommended for replacement of tube bundle, moreover, with more tubes plugged, gas oil cooling got affected in turn affecting H_2S absorption from sour gas. This sour gas was routed to the fuel gas network which was detrimental with respect to corrosion in the fuel gas network comprising pipes/fittings/burners/vessel, etc.
- **Troubleshooting:** It was found that the metallurgy of the tube bundle in the cooler was carbon steel. After thorough analysis, the inspection department argued that with gas oil containing H_2S that too at high temperature above 200°C stress corrosion cracking is possible in carbon steel. They recommended using a tube bundle of cupro-nickel metallurgy. Procurement action taken and after replacing the tube bundle as recommended, the problem got solved.
- **Lesson:** The unit was designed in and around 1973, and it might be that it was not covered in the design that time. It is now common that for cooling liquid hydrocarbon with H_2S concentration at higher % range and at high temperature above 200°C, cupro-nickel is recommended for use in tube bundles.

5.3.7 Frequent Leaking in Overhead Condensers Reducing Capacity in PDA Unit

- **Incident:** In PDA, propane vapour coming out from flash separators enters a series of condenser in parallel to condense propane which is subsequently recycled to extractor to produce deasphalted oil (DAO) for ultimate production of lubricating oil in other process units. These condensers' tube bundles leaked frequently thus affecting the unit t'put which in turn caused feed tank ullage problems and finally refinery t'put.
- **Troubleshooting:** The overhead propane vapour at about 70°C to 90°C ex DAO flash separator and at about 225°C ex asphalt flash separator were routed to a common header connected to multiple condensers in parallel. The metallurgy of the condensers tube bundle was admiralty-Brass. This unit was in close proximity to SDU where there were lots of chillers using ammonia as refrigerant and also there were ammonia leaks from these chillers and sometimes the smell of ammonia was felt even in PDA units also. It was doubted while maintenance crew opened the leaky tube bundle(s) for cleaning and plugging leaky tubes, the tube bundles might get exposure to ammonia by chance as naturally tube bundles laid open to atmosphere for a few days; the long exposure was anticipated during M&I shutdown. Hence, Tube bundles cleaning during M&I shutdown was done one after another or by engaging more groups of workers to avoid long exposure to the atmosphere because even if SDU was shut down during M&I, the smell of ammonia felt that time. After following these practices, performance of condensers were monitored but no such major improvement realized.

  Propane vapour from DAO flash separator and asphalt flash separator were collected for composition analysis. It was found that vapour ex asphalt flash separator contained H_2S at about 3%. Hence, doubt rose whether tubes were suffering stress corrosion cracking due to H_2S presence. The asphalt-propane mix furnace temperature was reduced from normal level of 225°C to 210°C, and the effect in the flash point of run down asphalt ex asphalt stripper was checked. It was found that asphalt run down sample flash point was as normal at above 200°C as earlier. The composition of propane vapour ex asphalt flash separator was carried out; this time propane content in the vapour came down from 3% to less than 1% which meant that asphalt was partially cracking at 225°C, which obviously produce H_2S and this cracking reduced appreciably at temperature 200°C; leak in tube bundles also got reduced subsequently by following this changed operating condition in asphalt-propane mix furnace; of course disciplined practice of condensers cleaning job were also followed as mentioned above.

- **Lesson:** It was not only one root cause, but two root causes which were affecting condensers life. So, in troubleshooting, we should focus to explore for more root causes if detectable or not.

5.3.8 Leakage from Thermo Well Point in Exchanger in DHDS Unit

- **Incident:** Hydrocarbon smoke observed from a thermo well point located at the outlet line of the heat exchanger. When approached to the location, it was seen that the leak was not from the thermo well, rather some hydrocarbon bubbles coming out from the thermo well pad. Unit operation became unsafe and shutdown of the unit was threatening to cause refinery shutdown as there was no ullage in the feed tank.
- **Troubleshooting:** As x-ray is not possible in the pad, only DP test is done after welding the pad as followed in general; however, pipe hydro testing should be done after fixing the pad. Project execution jobs records were investigated and found that fixing of the pad was carried out after pipe hydro testing. Also, a hydro test was done after fixing the thermo well nozzle and pad, but not sure whether tick welding of pad was done before hydrotest or not. It was also thought that there might be some entrapped grease inside the pad which was coming out as smoke through gaps in the pad

when temperature reached. Hence, instead of shutdown, operation continued with steam hose placed on the smoky zone. In every shift of operation, the hose was removed for a while and checked whether smoke disappeared or not. But after a few days, it was observed that smoke increased instead of stopping. Criticality of the feed tank ullage problem also got reduced by that time, and it was decided to take the unit shutdown. After the unit shutdown, re-welding was done for the thermo well to the main pipe, hydro test done, then pad welding done and again the pipe was hydro tested. This time there was no leak during operation.

- **Lesson:** For pad welding where recommended, it is better to carry out a hydro test twice as above.

5.3.9 Lube Oil Cooler Outlet Temperature High in NH_3 Compressor in Solvent Dewaxing Unit (SDU)

- **Incident:** There were two lube oil coolers with one in operation and other as standby. When lube oil temperature increased, cooler in operation switched over to stand by one, but even then, lube oil temperature remained high which caused a reduction in compressor load thereby reducing unit t'put.
- **Troubleshooting:** The twin coolers were interconnected through a 4 way changeover valve in the shell side (lube oil), i.e., by changing the position of the valve by 180°C, lube oil flow to the earlier cooler would be stopped and could be switched over to stand by one. As per design, if one cooler tube gets fouled up, then stand by one is taken in line. It was found that cooling water valves of stand by cooler were open. If the cooling water of the stand by cooler was kept isolated with its water in tube side drained off, then this high temperature immediately after taking it line should have not happened. But here, as the valves were open, cooling water flow continued in the idle cooler also causing it to get fouled up during course of time. For the time being, one compressor shut down out of two compressors to clean the lube oil coolers. In the next shutdown, modification was made to provide separate valves

to each cooler and to each side so that in case of both coolers fouling, two coolers can be kept in line simultaneously in parallel.

- **Lesson:** In addition to above, there should be a back flushing facility in the water side of each cooler and back flushing should be followed in regular intervals.

5.3.10 Reboiler Outlet Temperature to Stabilizer in CDU was Low Affecting LPG Recovery

Note: The problem also discussed in connection with column section 5.2.

- **Incident:** LPG yield on start-up after maintenance shutdown was on the lower side. It was observed that column top pressure, temperature and reflux flow/temperature were normal as was before shutdown and same crude mix processed after shutdown and stabilizer t'put also kept same. But bottom reboiler temperature was not increasing though reboiler hot stream, viz., gas oil CR flow and temperature were normal.
- **Troubleshooting:** Doubting reboiler temperature indication, crude tank with higher LPG potential was lined up, but yield did not improve. It was also found that RVP of bottom naphtha products increased which clearly indicated that there was LPG slippage occurring through bottom naphtha. It was also found that the bottom reboiler was cleaned during the shutdown. Functioning of a three-way temperature control valve in a gas oil CR (circulating reflux) line, i.e., at the hot fluid side of the reboiler checked which was also found normal. Effectiveness of cleaning of the reboiler was doubted as leakage was ruled out due to unchanged distillation property of bottom naphtha found. Hence, unit shutdown and bottom reboiler cleaned and the unit restarted, but the same problem remained. In the workshop, there was a spare tube bundle available with the same carbon steel metallurgy. It was thought there might be a permanent scale developed on running a reboiler during idling of the same in open air during shutdown. Hence, the unit shut down again and the reboiler was replaced with a spare

one as mentioned above which was also not new. Surprisingly, this time after startup, reboiler temperature increased with consequent improvement in LPG yield.

- **Lesson:** In troubleshooting no idea should be ignored, and it is learnt that there can be permanent scales developed on heat exchanger tubes due to unknown ambient conditions.

5.3.11 Higher Propane Coolers Outlet Temperature Affecting Extraction in PDA Unit

- **Incident:** Propane at about 48°C ex cooler/condensers is recycled to extractor to extract out DAO from top of extractor; the DAO-propane mix from top of extractor is then routed to a series of flash separator followed by steam stripper. This propane from DAO-propane recovery circuit joins the second stream of propane vapour coming out from asphalt-propane recovery circuit. The total propane then routed to the fin-fan cooler followed by water cooler/condenser and recycled back to extractor as mentioned above. The cooled propane temperature increased to about 55°C from normal level of 48°C, and thus it reduced extraction efficiency. To overcome the situation, external water pouring was done on the water cooler/condensers but it did not improve much. Ultimately t'put had to be reduced to reduce propane temperature but this was creating a feed tank ullage problem.
- **Troubleshooting:** Unit was shut down for tube cleaning of the condensers but not much fouling observed. After starting operation, propane temperature increased within 6 months of operation. Propane temperature at the fin-fan cooler was measured and found temperature was higher than normal. This time, the agency was lined up for external cleaning of fin-fans by hydro jetting with chemical-water mixture. After doing this, the problem got mostly solved, i.e., temperature reduced from 55°C to 50°C, and it lasted long. In the next shutdown, metal blades of fans were replaced with FRP blades and motors were also replaced with those of high rpm as it was observed that during stormy weather propane temperature got reduced by 1°C to 2°C; with this second action, propane temperature

further reduced to 48°C thus increasing extraction efficiency further and helped to increased t'put to design value.

- **Lesson:** Design vs. actual data comparison should be given first priority to understand before taking abrupt action like shutting down of the unit.

5.3.12 Product Run Down Temperature was Higher and Thus Became Unsafe to Route It to Storage Tank

- **Incident:** In Vacuum distillation units, the heavy distillate cooler outlet temperature increased from normal level of 80°C to about 95°C gradually. The product was routed to fixed roof tanks as per hydrocarbon classification. We know that any product run down temperature should not be close to 100°C as there is a chance of boiling in the tank if by chance there is water ingress in the tank which may be possible in the fixed roof tank by roof leakage if any and also possible in any tank with steam coil if any inside the tank; moreover, higher product temperature was affecting downstream unit where it was used as feed. This caused reduction in unit t'put.
- **Troubleshooting:** All design and actual data compared and found that the tubes of the rundown cooler must be fouled up as per calculation by following the method, viz., calculating actual overall heat transfer co-efficient and comparing it with design overall heat transfer co-efficient as explained in section 4.3. There was no scope of shutting down the unit as feed tank ullage was critical. There was no back flushing facility on the cooling water side. It was observed that the shell side of the cooler was insulated because the fluid in the shell side was of congealing nature. It was decided to remove the insulation. After removing the insulation, product run down temperature reduced by about 5°C. Also, a porous net-caging was provided on the shell side periphery to avoid human contact to the hot shell thus avoiding accident, if any. In the next shutdown, a back flushing facility was installed to avoid tube fouling earlier. This helped to maintain product run down temperature more or less steady during the operation cycle.

- **Lesson:** The above insulation removal concept was applied to all coolers in the plant along with provision of a back flushing facility in the cooling water side to avoid arising with such problems in other products.

5.3.13 In LPG Stabilizer in CDU, LPG Yield Came Down due to Lower Bottom Temperature

- **Incident:** It was like the above problem, but it was in the middle of the run, lower LPG production causing loss to the refinery.
- **Troubleshooting:** Like the previous problem, all operating data compared with design and previous cycle operation. Managing the situation by lowering stabilizer pressure was tried but, in that process, bottom temperature further came down and it did not solve the problem. Increase of top temperature was not possible as the overhead cooler was limiting. The bottom reboiler exchanger was stabilizer bottom vs. diesel circulating reflux and reboiler outlet temperature was controlled by a three-way control valve in diesel CR side. It was observed that diesel CR bypass flow was almost close as per control valve indication. Hence, fouling of the exchanger doubted, but cleaning the tubes of the exchanger meant hell of a job leading to shutdown of the unit. It was thought to reverse the opening of the control valve manually; when done, temperature started rising. Here, though people were understanding that minimum flow was through bypass, in fact it was reverse, rather maximum flow was through bypass; actually the controller was functioning in reverse direction which might be due to wrong work done by instrument technician on earlier job request, and it was difficult for operation guys from field indication. However, the problem got solved and LPG production normalized.
- **Lesson:** Knowledge on instrumentation by the operation guys quicken the problem solution process as this helps to understand which service areas assistance is needed in nick of time.

5.3.14 In CDU, Naphtha Run Down Temperature was Higher Than Its Flash Point Making It Unsafe for Operation

- **Incident:** Naphtha run down cooler outlet temperature was high and thus naphtha storage temperature in the tank became higher than its flash point. Though it was a floating roof tank, as temperature was higher by 5°C than flash point, t'put was reduced to keep temperature within control. This resulted in lower MS production and lower CDU t'put.
- **Troubleshooting:** Operating data compared with design; it was found that actual naphtha yield was higher than design, hence, operating CDU at design t'put means higher naphtha yield, but all exchangers in naphtha circuit were designed for lower flow; however, preheat train exchanger with naphtha service could take care with higher naphtha flow due to over design; similarly, naphtha run down cooler was also taking care but only during winter and monsoon period when cooling water supply temperature remained on the lower side. So, with normal fouling on the cooler tubes, it was unmanageable during summer. Hence, an external water shower was provided to reduce run down temperature. But this analysis resulted in revamping the size of run down coolers in future.
- **Lesson:** So, by comparing with design data, many a time troubleshooting completed without searching for unknown reasons in dark.

5.3.15 Fractionator Overhead Reflux Temperature High in VBU

- **Incident:** It was during the beginning of the operation cycle after M&I shutdown. Fractionators overhead cooler outlet temperature was high on the hydrocarbon side thereby increasing reflux temperature and in turn column top temperature increased; increasing reflux flow was limited by flooding in the top tray. This dictated in reduction in unit t'put and in turn resulted in critical ullage problems in feed tank.

- **Troubleshooting:** It was observed that cooling water inlet temperature was normal but outlet temperature was high but that happened after cleaning of tube bundles in M&I shutdown; it was not possible that tube bundles got fouled up so quickly; tube leakage, if any, was checked which may cause water ingress in reflux which in turn should increase in top temperature and top pressure, but there was no water in reflux, hence, tube leakage was ruled out. There was no water flow meter in the line to overhead twin coolers in series, only unit total water flow indication was there. It was doubted that whether maximum water was flowing through the header in the ground level for coolers in ground level and less flow was there to overhead coolers though the unit was designed for such a situation. This might be possible due to any entrapped air in the cooling water line at top most elevation; and it was found that this overhead cooler was at top most elevation of about 25 meters. The top most bleeder in the cooling water inlet line to that cooler was opened; immediately, air was gushing out instead of water. After about half an hour, air venting stopped and water started coming out. Then it was observed that top reflux temperature started to fall resulting in decreased column top temperature and the problem got solved. The bleeder kept open for about a day to ensure the water line got free of air. During M&I shutdown when the cooling tower sump emptied out, air ingress into the cooling water network was not totally vented out during commissioning of the cooling water system.
- **Lesson:** The top most coolers in all units should be identified and air venting should be done from these locations instead of venting from lower elevation during commissioning of the network.

5.3.16 Poor Heat Recovery in Heat Train in Naphtha Hydrotreater Unit (NHTU)

- **Incident:** In NHT, furnaces became limited due to increased firing of fuel at design t'put. The furnace stack made full open; burner registers also max open but draft was high as well as tube skin temperature was high. If firing reduced, furnace outlet temperature decreased thereby decreasing reaction temperature affecting desulphurization reaction which was not allowed and hence, t'put was reduced. But this resulted in lower MS production and feed tank ullage problems.

- **Troubleshooting:** It was observed that reactor effluent in the tube side could exchange heat transfer efficiently thereby decreasing feed–hydrogen mix outlet temperature resulting in furnace load increase. Initially this was not doubted as reactor effluent temperature in flash separator was normal; but when reactor effluent temperature at outlet of the heat exchange train was compared with design, it was found that heat recovery in the heat train got reduced, but with overdesign-margin in the downstream water cooler, ultimately flash separator temperature could be maintained. Hence, heat train consisting of 6 exchangers needed cleaning as there was no thermocouple in outlet of all exchangers, but for this not only shutdown but lining up agency, purging out of total high pressure system for handing over of exchangers became necessary which was a job for long period of about 15 days in total. It was observed that there was a DM water injection facility in the tube sides of exchangers, i.e., in the reactor effluent side. Operating manual studied, talked to licenser for DM water injection in tube side. But water injection in the running unit also required complete water draining from the flash separator as otherwise there was a chance of water ingress to downstream stripper. Hence, it was decided to take a bottled up shutdown of the unit for starting water injection in tube sides of the heat train which could avoid a lot of idling time of the unit.

  Before shutting down the unit, it was also planned to manage the operation for a few days by reducing the flash separator pressure marginally at least within limitation of allowable compressor suction pressure and discharge temperature. This was done as lower system pressure enabled to get desired hydrodesulphurization at reduced reaction temperature; but with this minor change, unit t'put could be increased by about 10% still less by another 10% from design t'put.

In the next shutdown, after DM water injection in tube sides of the heat train, heat exchange efficiency improved and the problem got solved.

- **Lesson:** Even after right troubleshooting, sometimes alternate innovation is required to manage the situation, if required.

5.3.17 In Diesel HCU, Exchanger Gasket Leaking in Every Shutdown

- **Incident:** Shutdown of the unit needs to be done in stages; first, feed in the second stage to be withdrawn after reducing its feed rate and reactor temperature. After stopping the second stage feed pump, fire caught in one exchanger with its service of feed in the shell side and reactor effluent in the tube side; it appeared that hydrogen circulation was not stopped due to obvious reason in line with procedures in operating mannual; thus there was flow of hydrogen in the shell side. The fire occurred by leak of the gasket of the exchanger; initially when leak started, smoke started coming from the gasket of the exchanger; sometimes it could be controlled by steam hose application on the gasket to douse the smoke, but sometimes it did not stop and Fire tender had to be called to extinguish; however, unit could be restarted simply by exchanger flange tightening and without handing over the exchanger to maintenance. But this recurred every time while withdrawing feed from the second stage of the unit.
- **Troubleshooting:** Generally, in an exchanger, first hot side fluid to be withdrawn then cold side fluid to be withdrawn; but in this case due to design as fixed by licensor, cold side fluid (feed to second stage) is to be withdrawn first though hydrogen flow is to be maintained in the shell side while hot fluid (second stage reactor effluent) is still flowing. The mass of hydrogen flow is inadequate to absorb heat of hot fluid and thus thermal shock as usual occurs. It was found that design considered such degree of thermal shock in the design of the exchanger though it was not working thus catching fire. Hence, maintenance was requested to investigate the problem. Maintenance found that the gasket was as per specification; so, they provided washers while tightening the studs in the

cover flange of the exchanger; this was found to ensure no leak from the flange during withdrawal of feed to the second stage, and the problem got solved.

- **Lesson:** Though it was an extempore solution, it is also a matter of improved practice and human art to reduce thermal shock while starting or shutting down the unit which should be documented as SOP (standard operating procedure). Also, after detecting any problem in the licensed unit, licensors should be kept posted which helps to get not only good management of change (MOC) from them but also to bring them in the loop of responsibility.

5.3.18 Frequent Leak in Furfural-Water Condenser and Downstream Line in FEU

- **Incident:** In FEU, the furfural drying column and furfural-water azeotrope distillation column overheads vapour join together and enters the condenser; the condenser outlet furfural-water mix liquid enters the furfural-water separator vessel where in one chamber furfural rich water and in other chamber water rich furfural phases are generated; the furfural rich water phase enters drying column and water rich furfural phase enters the azeotrope distillation column as mentioned above. There was sodium carbonate water solution dosing in the overhead of this system to avoid corrosion in the overhead and downstream circuit as mentioned above. It was observed that the overhead condenser and its downstream condensate line corroding during operation and compelled the unit to be shut down to repair the condenser and sometimes its outlet line also. This in turn caused lube oil production loss and FEU feed tank ullage constraint.
- **Troubleshooting:** The metallurgy of the condenser and its outlet line were of carbon steel. Sodium carbonate solution strength was religiously maintained to contain the pH from 6.5 to 7.5., but pH sometimes came down to level of 4.5 to 5.0; dosing pump was reciprocating type and sometimes pump gland leak caused loss of dosing rate. Gland leak was a perennial problem and could not be permanently resolved. It was also observed that while there was leak in the condenser

and its outlet line, there was no leak in downstream separator vessels as mentioned above, but sometimes leak observed in the water rich furfural line from that separator vessel bottom to pump suction. The exception of not leaking in that vessel was explored and found that there was refractory lining in that vessel that's why except vessels everywhere in this system there was leak.

One dosing pump was modified from gland packing to mechanical seal system; even then, leakage occurred.

It was also found while a solution of proper strength was prepared, strength came down sometimes in dosing pump suction and further downstream line. Solution was prepared from solid sodium carbonate by mixing with water and finally using agitator in the mixing vessel. It was found there was fault in the agitator but might be in the design of the mixing system.

Instead of exploring ideas to ensure preparation of homogeneous solutions, it was thought to explore the right metallurgy as this would be a foolproof solution of the problem. Monel metal was selected to replace the condenser and its downstream line, and the problem got solved. Sodium carbonate dosing was continued to take care of eventuality of any condensation and subsequent corrosion in overhead vapour line, in any carbon steel part inside the vessel and in the carbon steel outlet let of the vessel; this outlet line was not replaced as there was scope to change over to redundant line if any leak occurred.

- **Lesson:** The unit was designed long back when emphasis was given to reduce CAPEX as crude was very cheaper that time; but presently, there should be no compromise on investment for metallurgy as its contribution to overall CAPEX and OPEX are very low with respect to refinery profitability.

5.3.19 Tempered Water Cooler Leaking in VDU

- **Incident:** In VDU, heavy products like vacuum residue run down cooler, DM water was used as cooling water fluid in tube side as hot product temperature was high thus to avoid tube side clogging if circulating cooling water is used. The return

DM water is then cooled in a cooler where circulating cooling water was used in the tube side; the cooled DM water was then circulated back to the vacuum residue run down cooler. This DM water cooler was called a tempered water cooler. Even in this tempered water cooler, tube leakage occurred affecting availability of unit operation.

- **Troubleshooting:** DM water being neutral water should not cause tube corrosion as usually known; hence, DM water quality was checked and found okay. When the tempered water cooler was opened, it was observed that 5% tubes were clogged but these were in plugged condition. Though here hot side fluid temperature was not high enough, for e.g., only at about 60°C, fouling of tubes was not expected in a tempered water cooler, but on checking cooling water velocity, it was found low, so some clogging could not be ruled out. Due to this plugging, tempered water velocity in the cooler was high which may cause tube rupture but the higher velocity was within range and as we know that from this point of view, always 10% tube plugging is allowed. Tube thickness was checked and found to be decreased. That is why higher velocity tube leakage occurred. But why tube thinning? After going through the operating manual, it was found that there should be passivator injection in DM water in this tempered water circuit because DM water being of neutral pH try to slowly eat away carbon steel, that's why it is called as hungry water. But here no passivator was found to be injected.

  Hence, passivator injection was started in the tempered water vessel to get rid of the problem.
- **Lesson:** It was a silly mistake ignoring the possible impact of not using passivator particularly when there was no mention of the chemical's name as passivator.

5.3.20 In Delayed Coker Unit (DCU), Quench Column Overhead Fin-Fan Cooler High Temperatures

- **Incident:** High temperature in the fin-fan cooler was preventing condensation of the vapour thereby increasing loss to flare.

- **Troubleshooting:** In every cycle it was happening within two/three months of operation in the cycle. During M&I when the cooler was opened, dirt found but could be cleaned satisfactorily. There was no demister pad at the top of the quench absorber column. When the external surface was surveyed, a lot of sticky dirt was found. After online cleaning of the dirt, cooler temperature got reduced and the problem solved. It was found that by wind circulation, coke particles from nearby coke yards get deposited in all nearby fins. Hence, it was decided to carry out fin cleaning on a regular interval.
- **Lesson:** Nowadays, a closed storage of coke in domes like silo has come into effect as being practised by other refineries. This should be explored for implementation.

5.4 Pumps

5.4.1 Glycol-Water Solution Pump Not Discharging Adequate Flow in SDU

- **Incident:** In SDU, propylene was used as refrigerant to produce chilled propylene and passed through exchanger shell side where in tube side glycol-water solution having 30% strength flows with the help of a pump; the chilled glycol-water solution in turn flowed through series of chillers to facilitate crystallization of the fluid on the other side of the chillers before routing the chilled feed-solvent mix to filters used in the wax production process. Flow in the glycol-water solution circuit was controlled by flow and temperature controllers in the chillers. For temperature control, excess glycol-water solution flowed through bypass of the feed chillers. During course of time it was observed that bypass control valves gradually were reaching to close position when total glycol-water solution total flow in the circuit also was gradually decreasing; ultimately flow reduced to such extent that the above mentioned bypass control valves in the chillers' train got completely closed and chilling of the feed-solvent mix in the chillers' train got affected.

- **Troubleshooting:** Pump suction strainer was opened, but no dirt found; motor current was noted, that also was low as compared to design; pump actual flow was compared with design and found to be on the lower side. Stand by pump taken in line, but its performance also was not up to the mark and it could not discharge more than 80% as compared to requirement. Suction strainer of a stand by pump also opened, but no dirt was found. No load run of the motors were taken, current were normal; RPM of both motors were checked, both were normal. Pumps shaft freeness checked, found normal. It was decided to send both pumps to the workshop one after another; first the worst performing pump which was originally running was sent. It was observed that the impeller was about 80% eroded/corroded; it was surprising how this pump was delivering 70% of required flow; however, the impeller was replaced with spare available in the store and taken in line when flow became normal. A stand by pump also sent to the workshop; the same erosion/corrosion was observed on the impeller but it was less than the first one. It was also found that this pump was kept as stand by most of the time as it was giving 90% flow, i.e., less than the main pump in the beginning and thus the main pump was continuously kept running. This phenomenon indicated that it was not only corrosion but also erosion in the glycol-water solution system. The subject was studied in depth; it was found that metallurgy of the pump impellers were of carbon steel and no protective lining found. Hence, a feasible anti-corrosive solution was explored in the literature and the right inhibitor for such a system identified and it was available in the market also. After procuring and adding the same into glycol-water solution, the problem got solved.
- **Lesson:** It was a lapse in the design as well as a gap in fixing pH monitoring guidelines in the manual.

5.4.2 In FEU, CBD Pump Losing Suction Affecting Slop Recovery

- **Incident:** In FEU, a closed blow down (CBD) vessel was located inside an underground sump; slop recovery pump was located on the ground level for discharging furfural-water slop to the furfural drying column. The ground level pump was operating with suction lift instead of having suction head, but it was considered in design for pump net positive suction head (NPSH); but in actual operation, pump could not be started without priming and many a time it was losing suction. This caused CBD vessel ullage constraint and thus sometimes in desperation, content of CBD vessel drained partly to sump and the furfural-water from sump lifted by pneumatic pump and drained to ETP slowly, but in that there became BOD limit issue in the effluent water from ETP; so, this was not the desired solution. Pump suction strainer was checked; sometimes coke-like dirt was found, run after cleaning the strainer, but could not run on sustained basis. The whole suction line was flushed, but no solution was reached.

- **Troubleshooting:** The pump was sent to the workshop; the impeller was intact and no corrosion was found; no load current of the motor checked; current draw reported rather lower, motor shaft also found free rotating; pump's shaft freeness checked; found normal. NPSH calculation was carried out; as per calculation, the pump should run normal after priming. Reverse flow from the pump's discharge line was done to check discharge line thoroughness; found okay when both mechanical maintenance and electrical maintenance were about to give up, a person from mechanical maintenance asked to check the speed of the motor in no load condition. When checked, it was found that RPM was 1500 against requirement of 2900; surprisingly the different spec motor was exactly fitted to the foundation, i.e., it was of same size. There was neither tag marking nor any template on the motor; that's why it got mixed up while in the workshop. When enquired, it was found that the original motor got damaged long back and the same revived by rewinding and refurbishing by the motor vendor when template was lost; proper marking on the motor was done which got subsequently erased out. However, after changing the motor with the right one, the pump could run satisfactorily.

  In another case with the same pump problem recurred when all aspects checked as mentioned above, and thus there was no other alternative by open draining of CBD content and pumping as mentioned above but at a very slow rate to meet ETP criterion as mentioned above. It was decided to ensure extreme control in ingress of slop to the CBD vessel till shutdown when it was planned to open the whole suction line. When during shut down, it was opened it was found there was a layer of coke inside the suction pipe starting from the CBD outlet. It was due to ingress of furfural degraded coke which entered into the pipe and formed the layer during idling i.e., when there was no slop pumping thus reducing the effective diameter of suction pipe. This time the suction pipe diameter was increased from 4″ to 6″ as there was no easy solution to arrest degradation of furfural into coke. The problem got solved as observed during the next cycle of operation.

- **Lesson:** CBD vessel pumping is a problem which is observed in most of the process units and offsites, and in each case unique solutions have to be devised which is possible with sequential analysis.

5.4.3 Feed Flow Limiting after Revamp in SDU

- **Incident:** In SDU, capacity revamp was carried out; feed pumps were not modified/replaced as it was supposed to take care at higher t'put as per its characteristic curve. But actually, it was observed that the feed pump's discharge pressure was falling down and its downstream feed control valves to all chillers getting full open. This caused production loss on lubes and affected upstream units to operate at higher t'put.
- **Troubleshooting:** There was always line choking problem in SDU. That's why steam tracing and line temperature in both pumps suction and discharge lines were checked by field survey; but there was no doubt on line congealing. Downstream chillers circuit thoroughness checked by seeing individual flow control valves opening to respective chillers; control valves were floating; when feed flow was tried to be increased to design capacity, all these control valves were getting full open; the chillers' outlet temperatures were checked; there was no over-chilling in all these chillers. Chillers' downstream line up to filters inlet were checked, and found these were also through. Exchangers thoroughness

were also checked but not bypassed as it was not possible to get desired temperature with exchanger bypassed in actual operation; moreover, feed was going through shell side of the exchangers and total pressure drop through exchangers were 1.0 kg/cm^2 as per data book and also as per field pressure survey. When it was confirmed that there was no problem in the system, it was decided to look up again in the pump characteristic curve. It was found that as per characteristic curve, maximum feed flow achievable was much higher than revamp design requirement but differential pressure was just on the border line with rated impeller. It was also found from the characteristic curve that, using maximum impeller diameter, differential head increased by 10% but it reduced flow marginally and was found to be within limit of revamp capacity. The motor load cushion was also checked for operating with higher pressure while to achieve designed flow with differential pressure rise by 1.8 kg/cm^2 (10% increase from original 18 kg/cm^2); it was found that motor having that cushion. Hence, in one feed pump, maximum diameter impeller as available in store was used without trimming the impeller. The result was attractive and the problem got solved. Subsequently, the same change was implemented in the stand by pumps also. The pump characteristic curve vs. system curve (operating back pressure) are explained with varying impeller diameter as follows:

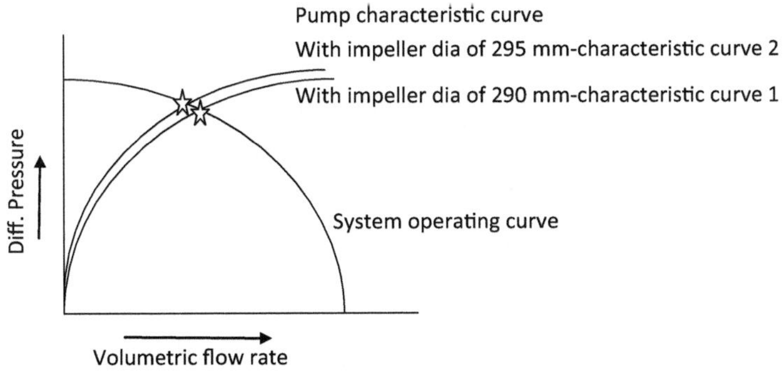

Figure 5.3 Determination of operating point of a pump using system curve and pump characteristic curve.

Note: the star marked point is the actual operating point indicating actual pump differential pressure and volumetric flow rate because where the characteristic curve and system operating curve is called actual operating point.

It is also seen that characteristic curve 2 with increased impeller dia has higher differential pressure with decrease in flow rate with respect to characteristic curve 1 as evident from the star mark shifting its location as above which meets the new equilibrium of operation after unit revamp which was not considered in the design by mistake.

- **Lesson:** 10% to 15% increase in pressure by reducing flow rate and vice versa, i.e., 10% to 15% increase in flow by reducing discharge pressure generally possible by using max dia impeller in 1st case in line with characteristic curve 2 as mentioned above and by trimming the rated impeller dia in second case in line with characteristic curve 1 as cited above.

5.4.4 Both Main and Standby Distillate Pump Mechanical Seal Leaks Causing Unit to Shut Down

- **Incident:** In VDU, 500N category lube distillate pump mechanical seal leak observed. As the pump fluid was of congealing nature, steam flushing was provided as a part of the sealing mechanism as selected by the pump supplier. As the leak increased, it became unsafe to continue the pump running; hence, stand by pump started, but that was not taking suction. To keep unit running without this distillate pumping, increasing draw rate of upper cut, light distillate and lower cut, heavy distillate was tried but that made qualities of those two distillates out of specifications as yield of 500N category, i.e., inter-mediate distillate was very high compared those two cuts thus unit could not be operated sacrificing inter-mediate distillate. Even, keeping the unit running by reducing unit t'put was unmanageable. This was leading to unit shutdown.
- **Troubleshooting:** The reason for stand by pump not taking suction was explored; its suction line from common header up to pump suction was checked whether congealed or not, but no chokage; its suction strainer was opened, found little dirt, cleaned, boxed up and again the pump started but in vain, it was not taking suction. The pump mechanical seal flushing mechanism was of conventional design, i.e., open circuit flushing with steam as the pump fluid was heavy, hot and of

congealing nature. It was observed that while steam flushing to mech seal was open, it should come out from seal vent hole in opposite direction, but nothing was coming out; it might happen that seal flushing steam injected might get condensed and thus no steam was coming out from vent; but while the other pump running with seal leak observed, it was seen that flushing steam injected was coming out from seal vent hole. Hence, it was doubted that in the idle pump, flushing steam might be sucked into the pump suction header as the pump was connected to the vacuum system. Hence, as a trial, steam flushing to mech seal to stand by pump was stopped, and the pump started. This time the pump started to take suction. People became happy, and pump mechanical seal performance watched, there was no leak, as steam flushing was must as per design, hence, then steam injection to seal started without further delay; pump operation this time got established without any seal leakage. Observing the operation for a few hours, the main pump which was having profuse seal leak was handed over to maintenance.

- **Lesson:** Steam flushing to idle stand by pump was open and the pump was kept idle for long thus with no flow of distillate on the other side and flushing steam flow from reverse continuously for a long time might have damaged the seal of stand by pump though as per design it should not be. Hence, next time onwards, practice of reducing steam flushing flow to idle pump and normalize it during running was started. However, the pump vendor was also informed about this.

5.4.5 In Sour Water Stripping Unit (SWS), Strong Sour Gas Smell in the Unit Causing Environment Threat

- **Incident:** In SWS, while making a visit to the plant, there was H_2S smell from inside the unit. When operators were asked, they could not reply as if they were habituated in it and this culture happens in operation where people avoid escalating the problem to higher ups and instead like in living with the problem till it reaches towards extreme situations. So, people were made to act in such situations. Maintenance crews also objected to work in that situation; finally the cause had to be discovered to keep unit running safe and also unit running was must to keep all upstream hydro processing units in operation.

- **Troubleshooting:** As there was no portable sensor to measure the extent of H_2S smell zone wise, a walk through audit was carried out to find out the leak source. The smell was vigorous in one zone of pumps, then walk through was limited in that area; each pump operation minutely observed. The feed pump with sour water as intake to the unit was found to have its mech seal vent kept open to the atmosphere, i.e., the seal vent connected with a hose and led to a nearby open funnel connected to line going up to ETP. It was also seen that the seal vent line was connected to a seal vent header which was leading to the incinerator of sulphur recovery unit (SRU); but the valve was closed and instead seal vent made open to the atmosphere as mentioned above. The maintenance supervisor was asked as to why such non-standard practice being followed. Came to know that there was back pressure from the seal vent header due to which seal venting was prevented and thus causing mechanical seal leak in that particular pump; that's why, they made this ad hoc remedy after which there was no seal leak. It was observed that this particular pump was located furthest from the incinerator and overhead elevation from the vent header was also high creating additional back pressure to the mechanical seal of that pump as compared to another pump. Maintenance was requested to lay a separate header of just 2″ dia for this pump's mech seal vent and to make this vent line joining the common header at an identified location which was close to the incinerator.

  After implementing the above mentioned modification, seal vent was routed to the close vent line and open venting was stopped. The pump operation without mech-seal leak was established.

- **Lesson:** Sometimes a culture develops to shortcut to solve the problem to avoid escalation to higher ups, and when the problem arises again, it becomes unmanageable. Senior executives in operation should survey units on a regular basis, oversee the operation whether carried out as per SOPs or not. Multidisciplinary brain storming at ground level on a regular basis can avoid such situations.

5.4.6 In CDU, Pump's Mechanical Seal Plan Not Working Causing Seal Leakage

- **Incident:** In CDU revamp, double seal provided in a pump with seal pot connected to its primary and secondary seal working on differential pressure mechanism following API Plan 53B; as per API Plan, external barrier fluid was given to upstream and downstream leg of seal pot pressurized by nitrogen as reference gas filled to a bladder placed inside the seal pot above the barrier fluid. Gas filling done once if gets lost, seal pot low-pressure alarm would appear, manually the seal pot again to be pressurized as otherwise seal pot level would go high tripping the pump; but pump should trip on seal pot high level without gas pressure becoming low indicating genuine leakage through secondary or primary seal. But in actual cases, there was no seal leakage, but seal pot low-pressure alarm appearing time to time causing interruption on pump operation for pressurizing the seal pot again and again or pump trip if there was delay in pressurizing the gas bladder.
- **Troubleshooting:** As per mechanism, if secondary seal leaks, then the seal pot's downstream sealing liquid level should increase and with a very high level in the seal pot, the pump should trip. But, in this case, there was a spurious pump trip due to the high level in the seal pot caused by the low-pressure alarm in the seal pot. It was found that reference gas was leaking out from joints at the top. It was thought that instead of batch pressurization with reference gas nitrogen, whether continuous pressurization from the inter gas line was possible or not. Inert gas was pure nitrogen with oxygen content less than 10 ppm and its header pressure was also matching the requirement of maintaining seal pot pressure; with this change, there should not be any low-pressure alarm on seal pot, and thus high level in the seal pot, if it occurs, should be

due to leak in the secondary seal of the pump. Pump vendor was discussed and modification implemented. Since then, the problem got solved.

A sketch on API Piping Plan 53B is shown as follows for understanding:

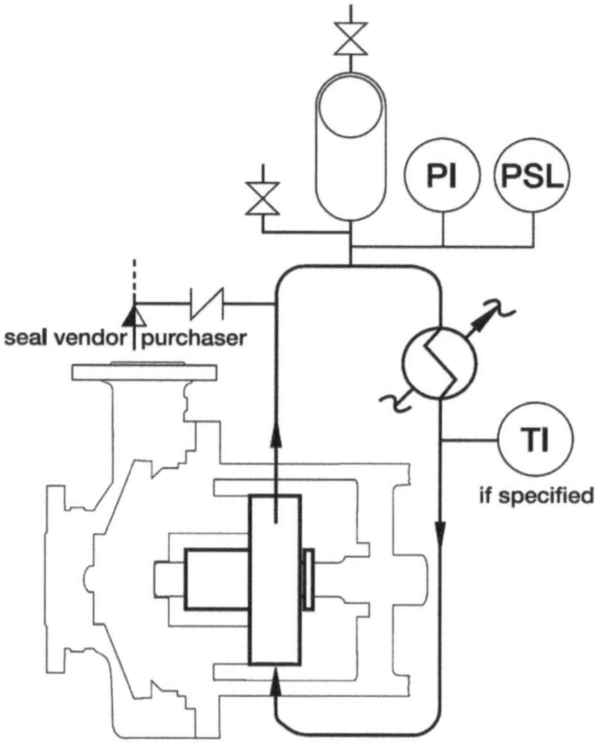

LEGEND: PI denotes pressure indicator; PSL denotes pressure switch low; TI denotes temperature indicator. Arrow symbol in the cooler denotes direction of cooling water flow.

Note: One level instrument with two level tapings to the above barrier fluid vessel should be connected though not shown in the sketch.

Figure 5.4 Sketch of API 53B piping plan for pump seal.

- **Lesson:** This was however an ad hoc solution not mentioned in API seal code; hence, for permanent solution, leak source should be identified which was carried out and original seal mechanism was re-established in future.

5.4.7 In HCU, Water Dosing to Fin-Fan Cooler Affected by Dosing Pump Gland Leak

- **Incident:** Water dosing to reactor effluent vapour inlet to fin-fan coolers were very important to avoid ammonium salt deposition in the coolers tubes thereby reducing flow of dosing water and ultimately choking the fin-fan tubes. But gland leakage of the dosing pump was causing less flow of water; change over to standby pumps and gland repair was regular phenomenon and gland leakage could not be rectified causing t'put reduction and production loss sometimes. Also, due to frequent gland leakage, sometimes both pumps kept running and thus there was no stand by pump for giving pump to maintenance. Moreover, this also was causing environmental problems in the unit due open draining of gland leak water.
- **Troubleshooting:** As the water was sour with presence of H_2S in the water, no gland found suitable to avoid its corrosion/ erosion as with gland sealing mechanism, it can never be leak-proof, rather there would be at least drop leaks; but due to allowing this drop passage of corrosive water through gland, gland getting corroded over time thus increasing gland leak. Hence, it was decided to replace the pump having gland packing system by pump having mech seal mechanism. To ensure stand by pump always available for maintenance, it was also decided to install new pump of higher capacity so that only one pump is sufficient for dosing; also, it was found that adjacent to existing pumps, there was space to install new pump; hence, parallel pump foundation job could be carried out including installation of new pump and in opportunity shutdown, hook up connection to new pump could be done. After getting the new pump of bigger capacity and with mech seal mechanism, one of the existing pumps could be dismantled for replacement with a spare new pump without shutdown of unit as hook up connection was also carried out for the spare pump. This resolved the important issue of water dosing to fin-fan coolers in HCU.
- **Lesson:** Compared to total CAPEX in HCU, investment cost for pump with mech seal instead of with gland packing is trifle,

and hence providing pump with gland package should not be used.

5.4.8 In PDA Unit, Asphalt Stripper Level Increasing Even with Spare Pump Running

- **Incident:** In PDA, through the extraction process with liquid propane as solvent, two phases are generated, viz., DAO-propane phase and asphalt-propane phase. Propane is recovered from these two phases from respective flash recovery systems and lastly from respective stripper columns to recycle the same propane for extraction subsequently. But the asphalt stripper level frequently went high due to its bottom could not discharge even with stand by pump running in parallel. This caused reduction in unit t'put resulting in "bright stock" lube production loss and feed tank ullage problem.
- **Troubleshooting:** Unlike DAO strippers, this problem occurred in asphalt strippers only. The bottom pump in asphalt stripper was selected as reciprocating steam driven due to high viscosity and congealing nature of asphalt. As there was every possibility of contamination with salt of stearic acid with consequence of creating foam in asphalt stripper, an antifoam dosing facility was provided to asphalt stripper inlet since commissioning and situation was manageable. This antifoam compound was developed in house with the help of interested chemical vendors. When level problems occurred, it was found many times that either antifoam dosing pumps were not working properly or the dosing line was choked. To avoid choking of dosing line, the total line was steam traced and also a check valve was provided in that line adjacent to injection point in the stripper feed line; also just at upstream of injection point, a bleeder was kept and line chokage, if any, was checked by opening the bleeder. Asphalt flash point also was checked and found normal and hence there was no doubt on any entrapped propane creating the pump vapour locking. Pumps' downstream line chokage, if any, was also explored, as run down asphalt was diverted to many routes like bitumen unit, visbreaker unit feed tank, and sometimes to fuel feed

tank of thermal power plant. With segregated flow keeping for a while in each run down line as mentioned above, line freeness was checked, and found all run down lines were clear against any choking. Even then, when stripper level problems occurred, maintenance was asked to check the pump, but they could not find any problem in the pump. Antifoam compound solution in dosing pot drained and fresh solution prepared, still there was a problem.

No procurement specification for antifoam compounds could be established except physical properties like density and flash point. Chemical was selected based on successful trials only. It was found that the chemical reached its expiry date. Hence, the chemical supplier was referred about the problem, and they replaced the chemical this time and the problem got solved.

- **Lesson:** Chemical supplier was asked to declare the composition of the antifoam compound exactly which was basically a silicone compound. After knowing the correct composition and strength of the chemical, in future, problems with the chemical got solved. However, earlier criterion of anti-foaming property was also kept in the specification.

5.4.9 In DHDS Unit, Rich Amine Pump Losing Suction Affecting Sour Gas Treatment in the Unit

- **Incident:** Sour gas was routed to the Amine wash column followed by the water wash column before routing it to recycle gas compressor in the unit. Lean amine entered the column from top from outside battery limit and rich amine from the bottom of the amine wash column was pumped and routed to the outside amine regeneration unit. After one idle shutdown when the unit was restarted it was found that, amine flow to the column if increased to desired value, rich amine bottom pump bottom pump could not discharge adequately resulting in column level increase. Lean amine flow meter checked and found to be okay. Also, it was found that recycle gas H_2S % increased. Hence, t'put was reduced to bring down recycled gas H_2S. This caused loss of diesel production.

- **Troubleshooting:** Pump changed over to stand by pump; problem remained. Pumps' suction strainer cleaned; no dirt found. It was observed that pump shut off pressure rose to normal, but after opening the discharge valve, flow did not rise to normal level and after some times, pump started to starving suction; pump could operate at low flow; motor checked on no load, no problem found. Pump taken to the workshop to check its internals, but no abnormality found. Rich amine discharge line in the offsite up to amine regeneration unit surveyed but no problem found. It was really puzzling. When the rich amine line all valves was checked with the help of a maintenance engineer, he suspected the opening of a rich amine battery limit valve; as per him, the valve was stuck up with very little opening for flow. To rectify the valve, unit shutdown became necessary which however would cause production loss. The engineer applied his mind. He suggested that he could rectify the valve without unit shutdown, i.e., by opening the bonnet cover of the valve and re-fixing the stem with valve seat groove which got loose. He also told if unfortunately at line pressure of 10 kg/cm^2 if the bonnet gland comes out, the unit should be shut down on an emergency basis and the rich amine pump should be stopped. We agreed with approval from competent authority with fire water tender keeping ready to douse the H_2S smell if such an incident occurred.

  But the maintenance engineer could successfully complete the job without any trouble to operation and thus rich amine flow established.

- **Lesson:** Sometimes innovative remedial measures may have calculated risk to execute for which other colleagues with expertise and engaging higher ups should be followed to take care of the probable risk. But formal HAZOP analysis should be done and approved before proceeding with any innovative method.

5.4.10 In CDU, Feed Pump Losing Suction at Revamped Capacity on Commissioning

- **Incident:** CDU capacity revamped by 10% with change in preheat train, feed pump impeller suction nozzle, booster pump after desalter; other modifications were not required as was not found necessary due to having cushion capacity on the rest. After commissioning, unit feed pumps found limiting at higher t'put thus enhanced capacity could not be reached.

- **Troubleshooting:** Feed pump suction strainer, motor checked; found no problem. Pumps taken to the workshop for internal checking; no abnormality observed. Pumps' discharge line thoroughness checked; no doubt. Suction pressure gauge was installed, found lower than design but could not be compared with previous data. Pump operated with level in crude tank; some improvement observed but not satisfactory. It was referred to the engineering service department. They confirmed that one inducer was provided in impeller suction to take care of lower NPSH due to higher flow and the same was not working alright. They came out with re-engineering of the inducer after which pump could discharge revamped capacity flow.
- **Lesson:** To take care of NPSH loss at higher flow, either suction line diameter to be increased which is a major job, or pump to be replaced with new one having lower NPSH requirement which requires long time procurement, pump foundation modification and sometimes motor change. Taking care by providing an inducer is very simple with least modification, but sometimes it may not click. Hence, for such type of innovative modification, one should be ready with a fallback arrangement like alternate inducer ready in hand with alternate engineering plan to avoid time loss in corrective engineering.

5.4.11 In FEU, the Common Stand by Pump Not Working with One Service

- **Incident:** In FEU, capacity revamp was carried out. A furfural flash column bottom pump and its spare were not revamped due to no change being required as per design. For furfural drying column bottom pump revamping carried out as not only capacity increase but also furfural ratio increase for extractor was envisaged in the design. All circuits were working alright, but when furfural drying column bottom pump due to any reason was taken out of operation, the flash column bottom spare pump was taken for this service as that pump was common spare; but the same could not operate at revamp capacity on drying column bottom pumping service. This reduced on-stream factor of the unit, i.e., sometimes leading to unit shutdown.

- **Troubleshooting:** As pump for flash column bottom service was not meant for cleaned fluid, i.e., with furfural-aromatic extract mix service as compared to clean fluid like pure furfural with drying column bottom, it was doubted that after shutdown, the common spare might not be working for drying column bottom pumping service but reason could not be established. Revamp design data sheet checked; it was found that with revamp, flash column bottom capacity did not increase much due to low-pressure operation with more furfural recovered and with only increase in extract load, pumps could take care due to having cushion capacity. In the case of furfural drying column service, as furfural ratio also changed in revamp design, drying column bottom pump flow increased due to which bottom pump was replaced, but in this process, the common standby pump's capacity with respect to bottom pump service might not be checked as it appeared. After revamping that common spare, the problem got solved.
- **Lesson:** Due to space constraints for carrying out revamp, sometimes the common spare pump concept is followed, but extra study should be undertaken to ensure that common pumps can perform satisfactorily for both services.

5.4.12 In VBU Gas Oil Absorber Bottom Pump Mechanical Seal Leak Frequently

- **Incident:** In VBU, this pump was provided with mech seal with seal flushing with self-flushing, i.e., one tapping from discharge connected with mechanical seal for seal flushing. The seal mechanism was in 1975 when the unit was commissioned. Seal leaking was frequent; maintenance also was used to replace it. This increased maintenance cost and sometimes caused unit availability problems.

- **Troubleshooting:** Initially the quality of the mechanical seal was doubted, but ruled out after changing the supply source. As the pump was at a temperature of about 250°C, hot alignment was recommended and executed, but the problem remained. Pump vibration measured but it was normal; fluid density and viscosity was compared with design; it was comparable. Maintenance was always complaining that fine dirt found in the seal while seal was meant for clean fluid service; after checking all the above points as mentioned above, it was decided that the dirt problem should be addressed. The circuit was checked that gas oil was flowing straight from the absorber column and sour gas was in contact with gas oil in the absorber. It was found that sour gas was wet and had H_2S and while purging out the gas, black particle was observed. It was understood that these black particles got absorbed in gas oil and while carried over, sometimes one/two particles might cause scratch in the seal surface. A strainer had been provided in a seal flushing line and the problem got solved. But due to lack of monitoring in strainer choking conditions, again in future seal leaked due to starvation in seal flushing.
  Finally the seal mechanism had been changed with API Plan 52 pump seal and the problem solved permanently. It may be mentioned that API Plan 52 is similar to API Plan 53B as explained earlier except here the seal pot top vent is open to the atmosphere with no inert gas bladder inside it.
- **Lesson:** Nowadays, all these problems are taken care of with modern API seal mechanisms and API 682 discusses all advancement in mechanical seals. Hence, refineries should switch over to these modern seal mechanisms for all pumps of old units.

5.4.13 In VDU, Gas Oil Pump Abnormal Vibration Started in Operation

- **Incident:** In VDU, this pump vibration was very high whenever started; when the standby pump started there was no vibration. The abnormal vibration also resulted in mechanical seal leak of the pump. This caused problems in availability of stand by pump and during failure of the main pump unit had to be shut down resulting in production loss.

- **Troubleshooting:** First of all, actual operating data compared with design when stand by pump was started, but there was no issue. Alignment of the stand by pump was checked, and found no problem. Suction and discharge lines supported the stand by pump checked; there was no looseness. Its suction strainer was opened, and found no dirt. Pump was taken to the workshop, no abnormality observed. It was thought that earlier there was no vibration in that pump; but this time vibration realized to be very high.

  It was observed that there given numbers of ad hoc support in the pump foundation; we know when the natural frequency of the pump including foundation matches with the dynamic frequency when the pump is in operation, resonance may occur resulting in vibration. To check against this, unit t'put reduced so that pump flow could be reduced; it was observed that vibration reduced to some extent indicating some shifting of actual frequency from natural frequency. The subject was referred to maintenance; they were asked why so many temporary inserts were given on pump base plate while fixing with foundation; they replied it was to adjust alignment; but this was against natural practice. It was also observed that the foundation was remade during previous shutdown as there was crack in earlier one which further damaged in operation. It was doubted that a new foundation was not made as per design. Hence, some further modification in foundation was done in consultation with pump vendors; pump vendors supported the idea of natural frequency vs. operating frequency clashing with each other. After modification as per advice of pump vendor, when pump again installed and taken in line, vibration reduced to large extent, and pump was operable.

- **Lesson:** There are several occasions of pump vibration with pump foundation design fault and it appeared since commissioning, but this time it was due to wrong workmanship of the maintenance crew. This was a heavy duty pump, hence, for heavy duty pump foundation job re-doing, the agency should be lined up under direction of the pump vendor.

5.4.14 In Lube Hydrofinishing Unit (HFU), Rich Amine Pump Vibrating and Gland Leak Continuously

- **Incident:** In lube HFU, the rich amine pump was of reciprocating type and provided with gland packing. As per design, inherent vibration was to take place in operation and to take care, a shock absorber was provided to protect against vibration due to oscillation. But in operation, the pump vibrated heavily. There was continuous gland leak loss which not only made amine loss but also made the area dirty. Pump foundation also got loose. This caused a unit availability problem for time to time sending pumps to the workshop for gland repacking. Also, connected lines stability became a point of question and panic.
- **Troubleshooting:** Suction and discharge lines support tightness strengthened and were tried one after another to check the status of vibration, but it worsened and the idea was given up. Suction strainers checked; no dirt found. It was also seen that actual flow was less than design. Design flow was higher due to higher amine flow considered in the design based on higher sulphur content in feed, but in actual, sulphur level was lower and with less amine flow, recycle gas H_2S level could be maintained thus controlling amine consumption. Amine flow increased to design level; vibration reduced; this not only increased the amine consumption even after gland leak repair which now holds on reasonable leak, but also operating crew from the amine supply unit raised the problem in maintaining its discharge pressure and causing amine supply problem to other areas.

  Hence, maintenance was asked to redesign the shock absorber which was of shorter duration than procuring new pumps. After replacing the shock absorber, vibration reduced and also gland leak reduced.
- **Lesson:** For every problem, there are alternate solutions. But, the pump should be replaced for a long-term solution as there was no scope of increasing amine flow in future due to no chance of feedstock quality change.

5.4.15 In Coker Unit, for Water Recovery from Coke-Water Sump, Water Recovery Pump is used but its Impeller Erosion Taking Place

- **Incident:** The pump takes suction from the sump and discharges to the API settler from where water was re-circulated by an HP pump to supply jet water to the coke chamber. The recovery pump as mentioned above found to discharge less flow during course of time.
- **Troubleshooting:** Suction strainer checked; coke particle found but not that much to justify loss of discharge. Pump after replacing the impeller found to perform normal for quite a long period and the old impeller found damaged on the edge. Inspection department reviewed and confirmed that it was not corrosion but erosion was taking place.

  Mesh size of the strainer increased which not only did not reduce flow but at the same time the pump was holding normal flow on a sustainable basis. Here, a small coke particle was eroding the impeller; after restricting it through high mesh size screen, impeller erosion reduced to large extent.
- **Lesson:** It was a simple solution; selection of the right size strainer in design may not be possible and to be optimized in operation only.

5.4.16 In HCU, Recycle Gas Compressor Tripping on Lube Oil Pump Failure

- **Incident:** In HCU, the recycle compressor is the heart of the unit, on its tripping, the whole unit gets shut down on trip security mechanisms to protect the catalyst and ensure unit safety. The compressor was provided with lube oil and seal oil pumps. Both lube oil and seal pumps were having standby with one on electrical drive and stand by on turbo drive. Medium pressure steam was used as motive fluid for the turbo driven pumps as lube and seal oil pumps were of very low capacities and 14 kg/cm^2 steam was adequate as per design.

  In both lube oil and seal oil services, motor driven pumps were kept on line, steam driven pumps were kept in warm up condition and in standby mode with auto start facility. Once during trip of motor driven pump in lube oil service as well as also in another occasion during trip of seal oil pump motor drive, the turbo driven pumps could not pick up adequate speed thus in both occasions, main compressor tripped on low lube oil pressure or low seal oil pressure in respective cases. These caused heavy production loss because shutdown of HCU results started and stabilization time of minimum 3 to 4 days.

- **Troubleshooting:** Maintenance crew were really concerned; they checked the drives and found no problem. As the turbo drive problem was common to lube oil and seal pumps, philosophy of operation was relooked, viz., whether turbo drive should be main pump or not with motor driven as stand by. On reviewing the manual it was found that the above idea should be followed. When explored why the exception was followed, it was discovered that the motor driven pump did not take start in time on lube oil/seal oil low header pressure. But this was a problem on the electrical side and they were asked to address the issue. This time the electrical department replaced the motors and modified the electrical relay circuit.

  After doing the above, next time onwards, turbo driven pumps kept running as main pumps and motor driven pumps kept on standby mode. This time, it was observed that on low header pressure, motor driven pumps took start without delay and took care of header pressure without time delay. Thus the problem got solved.
- **Lesson:** Problem occurred due to deviation from normal practice without approval from competent authority and that too without record which should not be followed.

5.4.17 In CDU, Crude Pump Casing Internal Scoring Take Place to Threat Repeat of Rupture

- **Incident:** Once there was catastrophic failure in CDU like rupture of crude booster pump at flash separator bottom. This caused a big fire in the unit affecting connected lines also which resulted in about 35 days shutdown for recovery work.

After starting the unit from this major shutdown, close watch was given to performance of these crude booster pumps as well as crude pump in offsite as casing internal scoring was also observed in that pump during shutdown. Once every week, one booster pump taken to the workshop kept standing in line. Same scoring observed again. Offsite pump scoring not found; however, thinking the scoring was due to recirculation in pump inside, the discharge pressure of offsite pump could be reduced by about 1.5 kg/cm^2 as there was cushion in reducing inlet feed battery limit pressure by compensating the lower discharge pressure by opening pressure control valve at feed inlet line inside unit battery limit. This idea worked and scoring in the offsite pump did not appear again. There was no scope to do the same technique in the booster pump at flash separator bottom as there was not much scope in opening of furnace pass flow control valves.

This led to problems in stability of operation keeping threat of further failure.

- **Troubleshooting:** There were three crude booster pumps with two in operation and one as standby when t'put was 100% it was also found that capacity of one pump covered 70% of unit capacity. So, when two pumps were running at 100% unit t'put, one pump run at 50%, i.e., just at turn down capacity of pump. When unit t'put reduced to 90% for the time being due to some reason like crude stock issue, the second pump was not stopped which caused each pump running below turn down capacity. Such operation for a single stage pump with high delta pressure like about 32 kg/cm^2 at about 180°C should not be safe. As such, now a day, two stage pump is given for such high delta P, high flow of about 290 m^3/h at high temperature. Hence, it was decided that unit should not run at 80 to 90% capacity; instead either it should be run at 100% or above with two pumps in line, otherwise, t'put should be brought down to 70% and one pump should be stopped thus the single would not operate below turn down capacity.

  After applying this philosophy, the problem got totally solved. However, for permanent solution, single stage pumps had been replaced by two stage pumps in future as discussed above.

- **Lesson:** It was due to the knowledge gap among operation crew and no past feedback from maintenance to operation crew that there was a tendency of internal scoring in pump casing. Operation and maintenance should continuously share their knowledge amongst each other through daily meeting basis with records of discussion on salient points.

5.4.18 In HCU, Fractionators Bottom Pump Losing Suction on Feed Filter Change

- **Incident:** In HCU, first stage reaction products after hydrogen recovery entered the fractionators; the fractionator bottom, i.e., unconverted, routed to the second stage reactor. The fractionator's bottom routed to pump via second stage feed filter. There was trip security on low pressure at this pump suction; hence, whenever delta *p* of the filter reaches the higher limit, it was changed over to the second filter to avoid tripping of the pump. But in this manual changeover process, the pump tripped on low suction pressure/low feed flow to the second stage resulting in tripping of the second stage feed pump with consequence to overall unit feed reduction to 70% of unit capacity. This incident was repeated causing production loss.
- **Troubleshooting:** As per the filter supplier, the filter to be operated with a maximum allowable delta p of 6.5 kg/cm^2 when at the same time pump suction reached just 1 kg/cm^2 above allowable minimum suction pressure. The filter then handed over to maintenance for cleaning and box up.

  In the changeover process, first inlet valve to stand by filter was opened, then to bleed out vapour, its top PSV bypass valve was opened and closed when the vent line was full hot and filter vessel also full hot; then outlet valve of the stand by filter was opened. Finally, the outlet valve of the main filter was closed and then its inlet valve was closed. But it was observed that this valve closing was not smooth and slow and heating of the stand by vessel was not complete.

  It was decided to change over the feed filter at 4.5 kg/cm^2 delta p, i.e., much below allowable maximum delta p to

continue operation, so that this filter changeover can be done very slowly keeping patience for filter filling and heating and isolating the main filter very slowly.
The problem got solved.

- **Lesson:** Here, technology/process was not violated but execution was not proper. This process is an art and experience helps to earn this art.

5.4.19 Feed Pump of Lube HFU Tripping on Start-Up High Current

- **Incident:** In HFU, feed pump was multistage centrifugal pump with suction pressure at about 2.5 kg/cm^{2}g taking suction from feed tank and discharge pressure of about 70 kg/cm^{2}g discharging feed to reaction system of about 60 kg/cm^{2}g pressure. Many a time, feed pumps tripped on start-up high current. After 2/3 trips, finally it got stabilized. After the first trip, a gap of 15 min was to be given and after the second trip, a gap of 30 min was to be given as per electrical department. This was causing unit start-up delay.
- **Troubleshooting:** It was observed, to ensure priming and heating of the pump, every time pump casing was filled and drained through casing bleeder to slop till pump is warmed up, but in this process, pump could not be heated adequately though priming could be done. It was also observed that feed tank temperature was about 85°C, and matching as per pumping temperature requirement as given in the pump data sheet. But taking a feel of suction line temperature, it was felt suction temperature was much less than 85°C, the whole suction line from tank up to pump suction was steam traced; it was found that in many places, steam tracer was not working due to trap malfunctioning. All trap bypasses opened; after some time, the suction line got totally hot. With this condition when the feed pump started. It did not trip on start-up high current and the problem got solved.
- **Lesson:** Again it was found that actual operating data was not compared with design data initially and this led to trouble in operation before the root cause was discovered.

5.4.20 Miscellaneous Troubleshooting in Pumps

Besides above, there were other troubleshooting experiences earned which are discussed below:

5.4.20.1 Pump motor burnt before tripping

One centrifugal pump motor burnt without tripping: it was observed that there was shaft jam in the pump either due to bearing seized or casing fluid congealed but coupling did not shear which made the motor to take higher load as possible; but even then, motor should have tripped before burning. One doubt was raised whether the motor's insulation value came down when it was idle though it should not be. It was decided that next time onwards, before starting this pump, motor insulation value will be checked. Simultaneously, the electrician checked the motor tripping current in the substation. When checked with the motor manual it was found that motor trip setting was wrongly kept at higher value. Hence, trip setting was reduced to avoid recurrence of such motor burning. Also, coupling with right strength should be used to ensure it sheers just above the tripping current value as this would provide extra protection to save the motor.

5.4.20.2 Pump not discharging though its shut off pressure was normal

There was another example of a pump not discharging though shut off pressure was found normal. Naturally, downstream lines thoroughness was checked and found that in one case, a check valve in the overhead line was found to be stuck up. In another case, the check valve at pump discharge found to be in reverse direction, in some cases, discharge valve gate found to be stuck up at close position detected with the help of maintenance.

5.4.20.3 Pump not taking suction with no restriction in suction line

In another case, the pump was not taking suction and when the suction strainer was opened; it was found that the strainer was positioned wrongly by a technician.

5.5 Compressors

5.5.1 In HCU Startup, Recycle Gas Compressor Suction Strainer Choking Frequently

- **Incident:** HCU start-up process initiated after month-long M&I shutdown. As per start-up sequence, when recycling gas compressor started, its flow gradually came down; on opening suction strainer, fine dust iron particles observed; strainer cleaned, again started, but again low flow problem and same dirt found in a suction strainer. This low flow problem and strainer choking happened repeatedly making delay in the unit start-up for a few days. People were puzzled about what to do.
- **Troubleshooting:** During maintenance shutdown, overhead fin-fan coolers were opened for cleaning; tubes were made of carbon steel and were kept open to atmosphere for about one month; also connected lines were of carbon steel and kept open to atmosphere; due to this atmospheric corrosion products generated which caused frequent choking of suction strainer. During shutdown, the fin-fan coolers with connected lines should have been kept under inert gas blanketing. However, when people felt direction less with such repeated suction strainer cleaning, compressor vendor was asked whether strainer mesh can be relaxed in size or not; vendor confirmed that this strainer was provided for commissioning of new unit; in normal operation, suction strainer can be removed unless the fine dirt affect the process. When the subject was discussed with the process licensor, they confirmed that the dirt trap catalyst in the reactor is supposed to take care of such quantum. The suction strainer design was with a net of holes of about 5 mm on which 40 mesh size strainers were provided. As permitted by the compressor vendor, this mesh of size 40 was removed keeping the strainer only with that big net in line. Also, however, a lot of fine dirt was cleaned out by this time.

  This time, the compressor ran without interruption and the problem got solved.

- **Lesson:** Carbon steel segments in the circuits should not be kept open to the atmosphere for a long time and compressor suction equipment should be under inert gas blanketing in shutdown.

  Also, the knowledge gap in strainer issues should have been resolved long back during unit commissioning.

5.5.2 In DHDT Unit, Makeup Gas Compressor Second Stage Discharge Pressure Not Rising

- **Incident:** It was the commissioning of a new unit. The makeup compressor when started it was observed that its first stage compression was as per design but in the second stage there was practically no compression. The stand by compressor was not yet handed over by the equipment supplier. This was causing delay in start-up.
- **Troubleshooting:** The loader valves were taken out, checked, no abnormality observed. A second stage suction strainer was opened, no dirt found; there was no restriction in flow from first stage discharge to second stage suction as otherwise, first stage discharge PSV would have popped. The loader/unloader operation sequence checked, found in order. A doubt was raised whether second stage discharge gas was leaking to first stage or not, but in that case, first stage suction temperature would have been very high and here that was not the case; moreover, second stage discharge temperature was low. Equipment vendors were attributing the problem on the process system like gas leaking out of the system from downstream circuits; but this was not acceptable when total system thoroughness was checked.

  Vendor was suggested to use the second stage loader valve of standby compressor in the main compressor and check. On doing this problem got solved.
- **Lesson:** In commissioning, the commissioning spare parts should be available to take care of such situations but here spares were not available. The equipment supplier should have been compelled to ensure availability, as such, in payment

terms, a good portion of payment should be linked to delivery of spares instead of keeping proportional amounts.

5.5.3 While Pre-commissioning DHDS Unit, Compressor Circuit Deblinding Done but Not Matching with Record

- **Incident:** When the operating crew reported that the deblinding job in the make-up compressor circuit was completed and the system was ready for hydrogen filling, pressurization, and running of the compressor as per commissioning guidelines, all blinding/deblinding jobs to be recorded in register with the countersignature of concerned operator/officer. Hence, it was asked to show the register; it was found that the compressor suction deblinding job was not countersigned, but the operators confirmed that it had been deblinded as per physical observation, but no one confirming who had got it done. It became puzzling to proceed for subsequent start-up activities or not when management pressure was felt regarding start-up schedule execution.
- **Troubleshooting:** It was decided not to take the risk of proceeding ahead without confirming. Hence, a suction flange was opened and found there was no material on the pipe mouth. As pipeline air blasting was done up to that flange to clean the line, there became a doubt whether under blasting pressure of 5 kg/cm^2, blind pieces can break and enter inside the line but that should not be possible as the piece was very thick and hard. Still officer was asked to try to see the suction line with a powerful torch. Wow! The person confirmed that a metal piece observed inside the suction line; it was decided to cut the line, but a trial was taken to bring it outside with the help of a long rod which became successful. It was also observed that the piece was not a standard blind plate, instead it was a thin sheet which was put as a blind in that flange, that's why it broke away and entered inside the suction line.
- **Lesson:** Nonstandard blind should not be used. It was once again established that accountability in following discipline/ guidelines in start-up/shutdown are extremely important

to avoid delay and hazard in the respective project. Also, one should not proceed for operation till a critical dispute is resolved.

5.5.4 In Wax Hydrofinishing Unit (WHFU), Make Up Gas Compressor Discharge Temperature Rising

- **Incident:** In WHFU, make up gas compressor was used to raise system pressure from 20 kg/cm^{2}g to 100 kg/cm^{2}g. Compressor discharge temperature rose and finally the compressor tripped. This was causing production loss of wax.
- **Troubleshooting:** Loader valves opened, found no problem; it was not like that discharge gas was short circuiting to suction because suction temperature found normal. Make up hydrogen purity was checked; it was as per design of 80% purity. Stand by compressor was taken in line; a similar problem occurred. There was no problem in suction flow and pressure. When it became really puzzling, the jacket cooling of the compressor was doubted; there was no flow meter on that line; simply utility water connection was there at one end, and at other end, the outlet line was open to drain. Jacket water flow in reverse direction was given to one compressor after stopping for back flushing. After doing this, discharge temperature improved but not up to the mark. Compressor supplier was consulted; in line with their advice, chemical cleaning was done in the compressor jacket; after this problem was finally solved. Similar exercise was executed in the standby compressor also.
- **Lesson:** As vendors did not recommend using DM water for jacket cooling, a flow meter was provided in the jacket cooling water line with back flushing facility so that when flow came down, immediately back flushing was given.

5.5.5 In Lube HFU, Make Up Gas Compressor Temperature Increasing

- **Incident:** In lube HFU, make up gas compressor was used to raise system pressure from 23 kg/cm^{2}g up to 65 kg/

cm^2g; there was a recycle compressor also which joined the makeup gas compressor discharge line. It was observed that makeup gas compressor discharge temperature was rising and tripping while there was no problem in recycling gas compressors. The standby makeup compressor was taken in line; similar problems occurred in that also. This was not only reducing lube production but also causing feed tank ullage problems.

- **Troubleshooting:** Makeup compressor suction temperature was also found increasing; loader/unloader valve operation sequence checked; no abnormality observed. Loader valves opened; valve seat/spring found damaged; maintenance replaced it; the same problem reappeared with new loader valves within a few days of operation. A doubt was raised on the quality of loader valves. Vendor was not convinced as earlier quality/specification was maintained as claimed by the vendor. Make up gas quality was analysed; it was found that H_2 purity was 70% as against design purity of 75% minimum. With lower purity, propane/butane carry over was taking place in H_2 make up gas; this propane/butane was getting condensed inside the compressor on compression thus through knocking effect it was damaging the discharge loader valves. Upstream H_2 supply unit was asked to improve the purity of makeup hydrogen gas; but to improve purity their production of MS was to be reduced, they reduced flash vessel temperature to some extent which marginally improved H_2 purity; they also increased the H_2 supply pressure by 1 kg/cm^2; on the lube hydrofinishing unit, liquid draining frequency from its battery limit knock out drum. This considerably reduced the compressor discharge temperature. Thus life of the loader valves improved to some extent.
- **Lesson:** It was an exigency phenomenon in upstream units, i.e., they were forced to extend the operation of CRU for MS production which could not be stopped as per plan.

  Improper production planning sometimes causes process/ equipment problems.

5.5.6 In Lube HFU, Compressor Crank Case Cover Glass Ruptured by Explosion

- **Incident:** In HFU, the make gas compressor which was a reciprocating type and took suction at about 23 kg/cm^2 and compress up to 65 kg/cm^2 suddenly exploded. Luckily the operator was not at that spot and got saved. When he reached, it was observed that its crank case cover glass ruptured, but the compressor did not trip. He stopped the compressor immediately, and isolated it by closing its suction and discharge valves. This caused non-availability of stand by compressor for a long period as the spare cover glass was not available. The unit subsequently suffered unplanned shutdown with failure of only one available compressor.
- **Troubleshooting:** The subject was not only of troubleshooting importance but also of safety/emergencies issues. There were generally two glands packing in the stuffing box between compressor cylinder housing and crank case. One was primary packing and the other was secondary packing. In between primary and secondary packing, there was a vent line without isolation valve so that any leakage gas from primary packing would be vented out to high point at atmosphere; if secondary gland packing also fails then this gas additionally would enter crank case and crank case naturally not designed to withstand high gas pressure; to avoid leakage through secondary gland, there was an inert gas connection to secondary gland packing to give back pressure to primary leakage gas thus leakage H_2 gas from primary packing and leakage nitrogen gas from secondary packing together should pass through vent line. There was an isolation valve in the inert gas line which was kept closed while the compressor kept idle for maintenance. It was found that the compressor after being handed over by maintenance for operation, all line up done but that inert gas valve was not opened which led to the explosion as mentioned above.

  It was decided to provide a differential pressure transmitter with an alarm of minimum 150 mmhg pressure so that the operator from the control room can observe and engage the field operator for corrective action. Nowadays, for enhanced safety, compressor automatic tripping on that low differential pressure is also included in the security system.

- **Lesson:** HAZOP review at regular intervals is necessary to establish more auto control and tripping facilities to prevent any failure on human error.

5.5.7 In NHTU, Make Up Gas Compressor Discharge Temperature High

- **Incident:** In NHTU, make up gas compressor took suction at 19 kg/cm^{2}g and compressed up to 33 kg/cm^{2}g to achieve reaction system pressure at 28 kg/cm^{2}g. This was as per design; but its discharge temperature was rising resulting in tripping of the compressor. This was repeated with consequent unit shutdown of not only NHTU but also downstream H_2 generation units (HGU) due to non-availability of desulphurized naphtha as feedstock for HGU.
- **Troubleshooting:** H_2 purity was 99% as per design. There was no short circuit of discharge gas to suction because there was no rise in suction temperature; the compressor body was also cool and as such there was jacket cooling in the design due to low compression ratio. Flow was also as per design. Loading/unloading sequence was checked and found normal, as such there was no problem in compression ratio and discharge pressure was as per design. Loader valve quality was as specified in the design. Hence, a doubt was raised with respect to the process or mechanical design of the compressor because this happened in commissioning of a new plant.

  As going to establish design faults can not solve the production crisis, another alternative was explored for. It was found that naphtha sulphur content was as low as 100 ppm as compared to design specification of 250 ppm maximum. Hence, it was thought that system pressure could be reduced because desulphurization reaction should be satisfactory even at lower pressure below 28 kg/cm^{2}g.

According to a trial basis, system pressure was reduced to 26 kg/cm^2g from 28 kg/cm^2g maintaining the same reaction temperature. This had given a benefit in reducing compressor discharge temperature as can be understood if the data are used in Eq. 4.2h under section 4.2. Also, there was no increase in sulphur level of desulphurized naphtha which remained unchanged due low feed sulphur as discussed above, and the problem got solved. As a procedure, the licensor was informed and they agreed for the change done as above.

- **Lesson:** Superior feedstock quality than design specification could take care of the design bottleneck of the compressor. The compressor design need not be revamped as it was reviewed and realized that there was no scope of feed stock sulphur level increase to 250 ppm, instead max up to 150 ppm in the long run; hence, no compressor replacement action was envisaged.

5.5.8 In CRU, Compressor Capacity Limiting Unit T'put

- **Incident:** In CRU, over a period of time recycle gas compressor flow needed to be increased to maintain mole ratio of H_2 and HC (hydrocarbon, i.e., unit t'put). But when the speed of the compressor increased to get higher flow, it was observed that its anti-surge control valve was getting open, which meant that compressor was getting back pressure from the system due to high pressure drop across the reaction system. The anti-surge controller was supplied by M/S CCC, Singapore and it was like black box to the refinery operating crew as it can only be checked by the instrument supplier, CCC; but in general, in no refinery, this CCC supplied instrument found to be with any problem. This problem was restricting unit t'put thereby losing MS production.
- **Troubleshooting:** It was observed that t'put reduced to 95% of design; but it was also observed from other equipment capacities, viz., furnace, pumps, column, reactor space velocity that these were having the capacity to increase unit t'put to 110% of design, but problem was only with the compressor as mentioned above.

Compressor datasheet was looked upon and compared with actual operating data. It was found that recycled gas molecular weight was considered as 8.5 whereas actual molecular weight even at mid of run condition was about 3.0. It was also observed that this molecular weight of 8.5 was only valid for compressors running with inert gas during regeneration, and wrongly this was considered as the basis for normal operation. To understand the effect of this mistake in design, let us reproduce the work done formula as shown in section 4.2 as follows:

$$W = P_i V_i/(\gamma - 1) \times ((P_f/P_i)^{1-1/\gamma} - 1)) \qquad (4.2g)$$

where W is work to be done by the compressor, P_i and V_i are suction pressure and volumetric flow rate at suction temperature, and P_f is the discharge pressure of the compressor. As the gas molecular weight decreases from 8.5 to 3.0, gas would be very lighter and thus volume would be higher; hence, the higher value of V_i in the above equation would result in higher work done requirement by the compressor.

Given this input to the compressor manufacturer, naturally compressor vendors should agree that here the compressor capacity is limited due to higher molecular weight considered, but there should be no limitation in turbine of compressor as the same deals with total weight flow rate of the gas.

Hence, it was taken up with a compressor manufacturer who reviewed the design with actual molecular weight to be considered and also to find out if any economic modification like changing of impeller keeping unchanged barrel was possible or not. Vendor came out with very low cost modification and took up the job to execute in next M&I shutdown.

After executing the same, unit t'put could be revamped to 110% of original design capacity as no other section in the system had any limitation.

- **Lesson:** Here troubleshooting to get back lost 5% unit t'put helped to jack up t'put additionally above 100%.

  Hence, aptitude for innovation gives a great mileage to productivity.

5.5.9 In PDA Unit, Compressor Breaks Down Frequently due to Liquid Carryover

- **Incident:** In PDA, compressors frequently break down on high discharge temperature; when the compressor opened by maintenance, loader valves found damaged; maintenance complained against liquid carryover which were causing frequent damage of loader valves as they experimented by changing the supply sources of loader valves also. In fact, there was continuous liquid carryover and the operator was supposed to drain condensate from second stage suction regularly; this liquid carryover was due to low purity of propane of 92.5% against design of 98%. Upstream units were requested to improve propane quality but they could not. This incident sometimes caused even shaft breaking of the compressor and the unit was without a standby compressor for about six months as the shaft was a long delivery item. This posed threat to unit availability.
- **Troubleshooting:** With thorough analysis it was confirmed that the maintenance argument was correct but at the same time it was observed that second stage suction temperature is also always high. Interstage cooler performance was checked; it was found that it was cooling inadequate. On exploring the condition of the cooler, it was found that more than 10% tubes, i.e., about 15% were plugged. When the inspection department was asked, they confirmed it was stress corrosion cracking and not due to fouling. On second stage suction gas analysis, about 1% H_2S was found and this was causing stress corrosion cracking. On literature study, it was found that asphalt-propane mix furnaces could be a source of H_2S generation even by minor thermal cracking of asphalt. The furnace outlet temperature was decreased from 225°C to 215°C and it was compensated by increase of stripping steam in asphalt stripper; with these actions, it was found that asphalt rundown flash point could be maintained and also H_2S content came down below 0.5%.

  The cooler replaced with new tube bundle and condensate draining from the knock out vessel was strengthened by maintaining a register on frequency of draining. In this way, the problem got solved. However, as a long-term solution, it is better to change over the cooler tube metallurgy to cupro-nickel.

Technical service department also took up the subject of debottlenecking the upstream propane recovery unit in CDU which also should be another permanent solution.

- **Lesson:** Here, in the absence of difficulty to solve root cause problems on an immediate basis, stringent monitoring on operation along with innovation with respect to interstage cooler performance solved the problem. Sometimes it happens that while exploring for one issue, other issues are also discovered and found to be related with the main problem.

5.5.10 In SDU, Second Ammonia Compressor (for Refrigeration) Tripping on Start

- **Incident:** In SDU, two ammonia compressors were to be in operation to achieve full t'put. Accordingly, in every startup after shutdown, initially the first compressor was started to get the desired chilling temperature for feed cut in; after the unit got stabilized with 50% unit t'put, a second compressor was required to start. The compressors were of multistage centrifugal type motor driven. As per procedure, before starting, the discharge valve of the compressor should remain closed; the suction valve should be partly opened to get desired suction pressure. Before starting the main motor, its lube oil and seal oil circuits to be established to get desired lube oil and seal oil pressure and temperature. The discharge valve made loose but closed. Then the main motor was started when gradually picking up current reached on achieving full speed of the motor. After the max pick up current fell back to normal current, its discharge valve quickly was to be opened, but in that process many times the compressor got tripped and the unit t'put increased got delayed.
- **Troubleshooting:** The problem was with the accuracy of the flow meter during start-up of the compressor. In actuality, even with partial opening of suction valve, full flow was achievable subject complete passage was available through discharge

valve, that is why before start-up max current got normalized, compressor sucked full load of gas and discharge valve crack open was enough to discharge that max flow which caused more time with max current thus tripping compressor. The electrical department was asked to review the max current timer setting whether allowable time could be increased or not, but they could not agree to relax the timer setting. Hence, it became the art of the operator to make a successful start-up, thus it was possible to start a second compressor with one or two particular sets of operators.

When the subject was critically analysed, it was doubted that the discharge check valve was passing which was giving back pressure to the compressor discharge end making the compressor tripped with high start-up current.

In the next shutdown, the check valves of both the compressors were replaced and also as a precaution, a soft starter was provided to the motor startup control circuit, and the problem got solved.

- **Lesson:** There was actually a gap in start-up current timer setting but both motor and compressors were supplied by foreign vendors long back and were difficult to make contact. On the other hand, a check valve may not always be leak proof. Hence, the modern concept of providing a soft starter gives added advantage in such cases, because this type of problem occurs in many units with HP motors where nowadays a similar concept is used to solve the problem.

5.5.11 In DCU, the Gas Compressor not Taking Start and Seal Oil Tank PSV Popped

- **Incident:** In DCU, there was a motor driven centrifugal compressor to take suction from fractionator overhead gas to recover LPG from off gas. During normal operation, once the compressor got tripped. On checking the power plant side, they confirmed that it was not on an overload trip; rather it was on process security trip. The process and instrument group could not diagnose the fault; when they gave clearance, it was started, but did not take start; rather it was found that its lube oil/seal oil common tank's PSV popped, and oil level in the tank depleted. The delay in starting the compressor caused flaring of off- gas and consequent loss of LPG production.

- **Troubleshooting:** From the process side, all actual operating data were compared with design; off gas quality also got analysed by laboratory; no abnormality found. Instrument maintenance was asked to check the security trip circuit; as per them, power tripping and seal oil low-pressure tripping appeared at the same time; hence, they were asked to check seal oil circuit; they could not find abnormality. It was also seen there was another parameter, differential pressure between seal oil and reference gas; based on this differential pressure controlled by fixed setting PCV, seal oil flows to different locations of the compressor as per design. It was doubted that this PCV was not functioning properly; the PCV was opened, taken to the workshop and its calibration was checked. It was found that the setting of PCV shifted from its normal position. As there was no spare PCV in stock, it was rectified and installed back in position. During malfunctioning of this PCV, reference gas short circuited and entered the oil tank, and with its pressure higher than tank PSV setting, it popped thus saving the tank from rupture.

  With the rectified setting of this PCV and topping up of lube oil in the tank, the compressor was started again and this time compressor did not trip and operation made steady.
- **Lesson:** There should be spare PCV in stock, and also there should be an indication on reference gas vs. Seal oil differential pressure with alarm to avoid repetition of the incident.

5.5.12 In Solvent Deoiling Unit (SDU), Refrigeration Compressor Tripping on Start

- **Incident:** In SDU, there were two motor driven positive displacement type compressors with in operation and other as stand by. One compressor was tripping on overload many times and the electrical department was engaged in troubleshooting. The problem was with the stand by compressor which was running without any problem, but this time, it was tripping on low suction pressure. This was causing delay in unit startup with the consequence of wax production loss.

- **Troubleshooting:** The standby compressor was checked by mechanical maintenance and instrument maintenance simultaneously. From mechanical side, loader valves were opened, found no abnormality; from instrument side, all other securities were checked whether tripping was taking place on other parameters also or not; instrument maintenance confirmed that as per tripping timer record, only low suction pressure was appearing first and others appearing after that which were consequence of low suction pressure tripping logically. Process side gas flow restriction possibility was checked; strainer opened; no dirt found; all line up was also correct.

  Ultimately, higher up executive in mechanical maintenance went to the position of the suction valve and doubted that the suction valve had not opened adequately and it appeared to be stuck up at crack open condition. The valve was dismantled from the position and doubt was confirmed. It was taken to the workshop for repair, but mechanical maintenance confirmed that it being a major repair, would take 3 to 4 days.

  Electrical department discussed the status of checking the electrical circuit of the second compressor; they confirmed that they found fault and rectified but the operating crew was not taking trial from fear that it would trip again. Then, in detail discussion with them as to what diagnosis was done each time after tripping; from discussion, it was transparent that this time, some fundamental root cause was identified and rectified. Their suggestion to take trial was accepted and the compressor was started. This time it was successful and unit start could be established.

- **Lesson:** Always there should be effort for exploring alternative scope in troubleshooting, and higher ups should intervene in crisis which helped to overcome the bias as mentioned above.

5.5.13 In HCU, Turbine of Recycle Gas Compressor Tripping Frequently

- **Incident:** In HCU, hydrogen recycle gas compressor drive was an HP steam turbine. To ensure tripping on over-speed, two tripping mechanisms were provided: one was mechanical lever, and the other was electronic over-speed governor. During one startup, it was observed that the compressor got tripped by a mechanical trip lever at a very low speed below allowable max speed of 11500 RPM. As the problem in the machine including the lever could not be identified, the machine was started again, but tripped after some time without notice by the operator. The frequent tripping was causing frequent emergency shutdown of the unit which not only caused production loss and feed tank ullage problem but also reducing catalyst life with respect to cycle length.
- **Troubleshooting:** Maintenance could not understand the defect in the lever; as such people doubted it as sabotage. CCTV camera was installed in that area. Next time, when tripped, camera record was seen, no abnormality was found. It was also fact that the machine when started crossed smoothly the lower critical speed limit of 5500 RPM with a just knock out sound which was usual but did not trip by the lever. When people stayed at the location at RPM of 7500, machine did not trip, but that time after operator left the place, machine tripped at about 8500 RPM. It was puzzling. Actually, mechanical trip lever was provided as a fall back measure to trip at RPM of 11700 if machine did not trip at 11500 RPM by electronic over-speed governor. Equipment manufacturer was informed about it; they were asked whether lever trip setting can be shifted from 11700 RPM to 11600 or not; they replied it was high-precision job and can be done at their workshop which was at a distance. Alternatively, they suggested that mechanical trip levers can be gagged, i.e., can be locked at site as electronic over-speed governor was already in line and as per practice in industry, nowadays mechanical trip lever is not used rather electronic over-speed governor only used which is more reliable. Also, in the control room, speed indication

was available with anti-surge tripping facility was also in line. Hence, the mechanical lever was kept locked, and the machine started with only an electronic over-speed governor in line. To check the efficacy of the governor, once the machine deliberately got tripped at low speed before feed cut in by changing the trip setting of the governor. It worked well. Then the setting normalized and machine restarted and feed cut in done as per procedure. This time there was no tripping and unit operation got stabilized.

- **Lesson:** Troubleshooting became a unique example as it could not be understood, when maintenance never touched the lever for checking, how from its own it was malfunctioning. Hence, they should send the lever to the equipment supplier to know the reason.

5.5.14 In HCU, Fractionator Overhead Gas Compressor Tripping

- **Incident:** In HCU, fractionator overhead gas was cooled/ condensed through fin-fan cooler followed by cooler/ condenser and gas/liquid mixture then routed to accumulator vessel. The vapour from accumulator then routed to suction of a three-stage reciprocating motor driven compressor to recover LPG and residue light naphtha from it. In between the stages of compression, there were intercooler to condense the LPG and residue light naphtha, if any. But compressor third stage discharge temperature most of the time went higher and compressor tripped. To avoid tripping, compressor load was required to be reduced and hence, standby compressor also had to be taken in line. This increased power consumption and energy loss. Also, when standby compressor was not available, flaring was to be done from suction to avoid third stage discharge temperature going high and compressor tripping. This reduced LPG recovery and production loss of LPG.
- **Troubleshooting:** As per design, first stage suction pressure should be 1.0 kg/cm^{2}g, but in actual it was about 0.5 to 0.6 kg/cm^{2}g; this however was manageable as far as first stage

and second stage were concerned but, in that process, these two stages could not raise second stage discharge pressure to maintain required suction pressure in third stage; thus third stage discharge temperature went high up to trip setting. Maintenance also complained of compressor loader valves damage and doubted on liquid carry over. Hence, condensate draining from inter-stages was reviewed by checking inter-stage knock out vessel level control valves and liquid pumps functioning. After strengthening these two controls, liquid carry over was prevented to a large extent, but even then, the compressor was tripping on third stage high discharge temperature.

Hence, the root cause of the problem was first stage low suction pressure. It was found that delta pressure between flash zone and column top was higher than design. The delta pressure across overhead cooler/condensers was even marginally higher than design though this section was revamped by adding condensers. It was found by calculation that though additional condensers were given, pressure drop across vapour line and condensate lines were higher by about 0.2 kg/cm^2; moreover, modification in this circuit was time-consuming job as there were a lot of branch lines at inlet and outlet of many condensers. On the contrary, when it was looked upon modifying column internal from trays to packing in some sections, it appeared to be more attractive from execution point of view, i.e., less time required. There was also problem in the kerosene zone where tray modification was done during previous revamp shutdown; when this corrective job in any case to be carried out in next shutdown, it was decided to replace this section trays with structured packing which also indicated that pressure drop in column should come down by 0.2 kg/cm^2; hence, the same carried out in next shutdown which not only reduced pressure drop but more than this by avoiding occasional flooding in that section.

The problem got solved after implementing the above.

- **Lesson:** This was basically corrective engineering after revamp shutdown, and this usually happened in most of the revamp projects. Hence, operating engineers should engage themselves with designers or follow more stringent coordination with designers to avoid such problems.

5.5.15 In HCU, Makeup H_2 Gas Compressor Tripping Frequently

- **Incident:** In HCU, there were three makeup H_2 gas compressors with two running and one kept as stand by. Each compressor was three-stage reciprocating and with HP motor drive; the compressors compressed from 20 kg/cm^{2}g at first stage suction up to 180 kg/cm^{2}g at third stage discharge. It was observed that these compressors were tripping at third stage discharge high temperature frequently and to avoid this, unit t'put was to be kept lower than 100%. Also, compressors loader valves are also required to be changed frequently. This caused an availability issue of the compressors resulting in unit availability and production loss and feed tank ullage problem. Also, this increased cost of maintenance.
- **Troubleshooting:** Actual operating data were compared with design; there was no abnormality. There was no question of condensate carry over with H_2 and H_2 purity was also as per design, i.e., at 99%. The impurities were compared with design and found to be within allowable limits. The inter coolers performance was checked; it was observed sometimes third stage suction temperature was high. The intercoolers were opened, there was no abnormal fouling, but there was some improvement after back flushing of the condenser tubes; but this improvement did not continue for a long time. As a part of a development project, one compressor was modified with step less control as originally the compressors were with loading facility at 25%, 50%, 75%, and 100%. With implementation of above and two compressors running including this modified compressor, unit max t'put could be maintained, but there was availability problem with this modified compressor also due to its loader valves getting damaged like other ones. Hence, loader valves modifications were looked into, but loader valves problems remained even after changing with the new

design of loader valves, and sometimes problems with bearing house.

Hence, there came doubt on maintaining all parameters of compressors from process up to machine parameters like maintaining lube oil pressure/temperature, vibration/temperature of bearings, motor winding temperature, etc.; to ensure religious monitoring, on line continuous monitoring mechanism was explored and after discussion with prospective bidders tender was floated. After implementing on line condition monitoring, the problem mostly resolved as this helped in taking on time action on the compressor circuit before it reached to break down.

- **Lesson:** This was an example of an inherent problem with older design, and sometimes even without finding a major gap in operation, troubleshooting achieved through taking up development projects like this.

5.5.16 In Cat Dewaxing Unit (CIDU), Make-Up Gas Compressor Tripping on Low Suction Pressure

- **Incident:** After the debottleneck shutdown unit restart activities started; make-up hydrogen gas compressor which was 3 stage reciprocating motor driven compressor was started but compressor was tripping on first stage suction low pressure causing delay in unit startup.
- **Troubleshooting:** Suction strainer was cleaned but nothing found; the compressor went to maintenance during shutdown and hence, maintenance was discussed on its performance; its discharge temperatures were normal, as such compressor performance could not be checked as on starting, it got tripped immediately.

  While checking the lines, it was found that there was a spill back line from first stage discharge to first stage suction through a restriction orifice; also, in that line there was an isolation valve which was found to be in closed position. The problem got solved after opening the valve.
- **Lesson:** Operating crew should check all line ups of the compressor before starting it.

5.5.17 Inert Gas Compressor of SDU Tripping on High Discharge Temperature

- **Incident:** It was a single stage centrifugal reciprocating motor driven compressor. During the course of operation, its discharge temperature was running high thus limiting its capacity and sometimes tripping thereby causing LOBS production loss.
- **Troubleshooting:** Discharge downstream back pressure found normal; suction pressure and flow also was normal. There was no standby and both the compressors were in line to get full t'put as per design. Hence, in the next opportunity to shutdown the unit, the compressor internal was opened and found that there was black deposit in the impellers and casing which was causing high discharge temperature. This deposit occurred due to time to time liquid carry over to compressor suction by upstream process upset; compressor suction knock out vessel demister also was checked and found damaged which was also rectified along with cleaning of compressor internals. The problem found solved in the next startup.
- **Lesson:** It was a cascading effect of one problem in one section of unit to another section; hence, extra care is needed to avoid knock out pot level going high by not wasting time in reducing unit t'put if level appears in suction knock out pot.

5.5.18 Anti-Surge Control Valve of Inert Gas Compressor in SDU was Getting Open

- **Incident:** It was observed that the anti-surge control valve was 30% open and there was noise in the compressor. Actually, the compressor should be operated at design capacity with an anti-surge control valve in close position. This forced us to reduce the capacity of the compressor but even then, noise did not reduce causing stopping of compressors and shutting down the unit.
- **Troubleshooting:** Before stopping, a quick survey in the unit was done; it was found that downstream chillers' tube side through which inert gas was flowing was choked as

found by delta pressure across the chillers. The tube side was immediately bypassed and the chillers' tube side got dechoked by stopping refrigeration in it and instead starting hot refrigerant flowing through the tube side. After dechoking the tube side, the gas flow established through chillers and the problem got solved.

- **Lesson:** This was a common problem in SDU in many chillers sections and hence, avoiding water ingress into the system in SDU is very important.

5.5.19 Air and Nitrogen Compressor in CCRU Mechanical Seal Got Leak

- **Incident:** These compressors were non-lubricated with dry seal as per process requirement. Hence, to ensure total prevention of any dirt entering into the seal, acid pickling of suction lines were done before commissioning and there was no problem in operation. But after one year of operation such leaks increased the cost of production.
- **Troubleshooting:** On exploring records, it was found that these compressors went to maintenance in previous shutdown and suction lines were kept open to the atmosphere when rust might have formed. Actually, these two suction lines should have been kept sealed preventing exposure to air. Again acid pickling was done in these two lines and the problem got solved after replacing seals.
- **Lesson:** Sometimes it is observed many practices followed during commissioning are ignored subsequently due to inadequate knowledge transfer. Hence, while preparing SOP, design features, licensors and OEM's instructions and additional precautions which had to be taken during commissioning should be taken care of.

5.5.20 NASH Compressor in SDU for Wax Production Tripping Frequently

- **Incident:** NASH compressor was a water ring vacuum compressor which was of positive displacement type; it was motor driven. Inert gas, nitrogen was used as fluid and the compressor was supposed to create vacuum in the filters section at the upstream of the compressor suction. The name 'NASH' was given after the name of the compressor manufacturing company. It was observed that the compressor was tripping not on an electrical circuit but on process logics. But from trend graphs it was difficult to find the root cause whether it was low suction pressure or low water flow which was used for forming a water ring inside the barrel of the compressor as per compressor design technology.

- **Troubleshooting:** As minute differences in time in the trend graph could not be confirmed, it was decided to stand near the compressor suction and water pump to identify which was the root cause. It was observed that compressor discharge knock out pot water which was supposed to go back to water pump suction, sometimes get affected by delayed action of knock out pot level control valve; this created water flow starvation to water pump suction. The control valve got rectified and the problem got solved.
- **Lesson:** Lag in instruments response is one of the most important parameters in process control.

5.6 Utilities and Offsites

5.6.1 In SRU, Low-Pressure Condensate Vessel Venting Steam Rate Very High

- **Incident:** For condensate recovery, there was a flash vessel inside the unit; it was mainly low-pressure condensate which led to this vessel where un-condensate steam was vented out and condensate from the vessel bottom was routed to power plant condensate station. It was observed that the amount of venting steam was very high which was not only creating noise but also making the area unsafe and dirty due to condensate drops showering from the venting steam.
- **Troubleshooting:** It was found that total condensate load in the upstream equipment was not higher with respect to

design, but the problem was that total condensation was not taking place; instead the amount of un-condensate was very high. On analysing the steam heater load, it was found that actual sour gas flow was very less and steam heater control valves were not functioning properly with such less load though process sections were capable to operate at such load just above turn down load.

Instrument maintenance confirmed that control valve modification may cause problems in operation with higher load, if necessary. They suggested two types of control valves in position with smaller one to operate at lower load and bigger one to operate at higher load; but this was a matter of long-term solution.

It was thought what can be the short-term solution. The idea of adding DM water in the vessel inlet line was thought; the limitation of the vessel downstream line was checked; found to be in order. Hence, with a small modification, one piping connection was given from unit DM water line to vessel in let-down condensate line; this reduced steam venting and the problem of noise and condensate drops dripping from venting steam stopped.

- **Lesson:** This was a case of alternate ideas for survival in operation.

5.6.2 In PDA Unit, Low-Pressure Condensate Vessel Safety Popping

- **Incident:** In the PDA unit, after ROSE technology by the licensor, Kellog, USA was implemented, additional medium pressure steam (14 kg/cm^{2}g) was required for the new ROSE heater. The condensate from the exchanger was taken to a vessel where from top low-pressure steam generated (4 kg/cm^{2}g) and from the bottom medium pressure condensate routed to second stage vessel where from top atmospheric pressure steam vented out and from the bottom low-pressure condensate routed to condensate network under pressure of this second stage vessel. It was observed that PSV of the second stage condensate vessel was popping.

- **Troubleshooting:** While pressure of the second stage vessel was monitored it was observed that PSV was not malfunctioning, in actual the pressure of the vessel was higher. The opening of the bottom control valve of the first stage vessel was checked to see whether live steam was passing through it or not. The delta *P* of the control valve was also checked to see whether it was lower than design or not; no abnormality found. Then control valve opening was manually reduced; it was observed that PSV popping did not stop, rather first stage vessel level was increasing.

  To maintain the pressure of the second stage vessel to facilitate flow of condensate from its bottom, a pressure equalizer line from first stage vessel top to second stage was given in design with line size of 1″ dia without any control valve in the line, but with a gate valve. As per design, the steam flow from 1″ line would not be higher to increase the pressure of the second stage vessel beyond its PSV set pressure but instead it would pressurize enough to push condensate from its bottom. As there was no other reason found against PSV popping, the gate valve in 1″ steam line as discussed above was also pinched to reduce the pressure of the second stage vessel. After doing this, PSV popping stopped.

  Also it was decided to install a control valve in that equalization line to control the pressure of the second stage vessel for long-term solution.

- **Lesson:** This was a case of a gap in the design, and operation department should engage themselves during design by third parties.

5.6.3 In PDA Unit, Low-Pressure Condensate Line Hammering and Leaking

- **Incident:** In the PDA unit, low-pressure steam was used in extractors top section heaters, DAO-Propane flash separator feed heaters and pipeline heat tracing. The condensate from all these equipment and pipelines entered the condensate header through a steam trap and routed to power and utilities plants. The condensate header inside the unit was always making noise like hammering sound and consequently causing leaks through pipeline gasket joints and making the area unapproachable due to condensate shower from header at the pipe rack.

- **Troubleshooting:** As the number of steam heaters were limited, say, 8, it was easy to check their outlet condensate traps to see whether these were working or not; as such all were working. But there were huge no. of steam traps in pipelines' steam tracing, checking of which was a herculean task. Every time after condensate header leaking, some operating crew reported that they identified some malfunctioning traps which were isolated before commissioning the condensate header after leak rectification. But in vain; the line hammering restarted and leak also reoccurred after a few days of operation.

  P&IDs also were not updated and documented properly. Hence, a task force was formed involving 2 engineers to oversee the checking of all traps by marking the location of checking and everyday reporting on the status. After a few days of disciplined checking, all malfunctioning traps identified, new traps resourced by maintenance as otherwise a lot of steam venting would take place during operation. Unit shutdown; all faulty traps replaced by new ones; unit restarted with condensate open draining and without commissioning of the new traps. When no hammering was felt in the condensate header, the new traps were commissioned one after another when also no hammering was felt from the header.

- **Lesson:** If the trap malfunctioning was attended with minimum sound at any location of the header at the first opportunity, then this problem would have not come to this level.

  Hence, discipline of attending even all minor jobs in the beginning to avoid it becoming a major problem in future making it complicated to address as mentioned above.

5.6.4 Sour Water Line Leak Repeating

- **Incident:** In HCU sour water line ex hydrocarbon water separator vessel found leaking. As water could not be drained to open, the unit had to be shut down and restarted after line leak repair. But this leak reoccurred after a few months of operation. This caused production loss.

- **Troubleshooting:** This time, Inspection checked the metallurgy and class of the affected pipe; it was found that it was of carbon steel of normal grade instead of NASH grade which should be used in sour water lines. After replacing the affected pipe with a new one of NASH grade, recurring pipe leak stopped.
- **Lesson:** This was due to a gap in quality control in project execution. Imposing a penalty on a contract does not solve the problem, but performance evaluation which is not done presently should be done for all contracts and should be given a weight-age in the next contract.

5.6.5 Refinery Instrument Air Pressure Came Down Causing Total Refinery Shutdown

- **Incident:** It happens many times in the refinery that instrument air header pressure alarmingly comes down, challenging continuation of operation and posing threat towards total refinery shutdown. In fact, once a refinery shutdown owing to low instrument air pressure could not be avoided.
- **Troubleshooting:** In one instance, it was observed that this situation occurred in a new unit for which a separate instrument air vessel was provided to ensure time for safe shutdown of the unit in case there was a problem in the refinery instrument air network. The vessel supply air pressure was normal, the PSV of the vessel was also functioning; there was a pressure control valve in the instrument airline at the inlet of the vessel which was also found in more open position when vessel pressure came down; this helped to get back vessel pressure and low-pressure alarm went off and people left the place; but again after some time vessel low-pressure alarm came and by the time people reach the location, unit got shut down on instrument air low pressure.

The control valve action was studied without starting the unit. It was observed that the control valve suddenly got close of its own. The control valve was taken to the workshop, rectified and reinstalled and the problem got solved.

In another instance, refinery instrument air header pressure came down to such extent that a standby air compressor had to be started; but the compressor being of HP motor drive caused more power consumption making energy loss particularly when leak source identified and rectified. Stopping of the stand by compressor caused reliability of maintaining air pressure. The problem got addressed by installing an auto loading facility in the stand by compressor thus avoided un-necessary power consumption when there was no loss of air.

- **Lesson:** The instrument air compressor is one of the supercritical equipment in the refinery. Hence, categorization of equipment with respect to improving equipment reliability is most important to address.

5.6.6 HP Steam Temperature to Turbine of Recycle Gas Compressor (RGC) of HCU was Falling

- **Incident:** In refinery, HP steam generating source, the power plant was located at about 2 km from HCU. In the HP steam network, the other sources like coke calcination unit (CCU) and hydrogen generation unit (HGU) were there to supply HP steam in the network but CCU also was located farthest, i.e., at about 2.5 km with only HGU was in close proximity of HCU. The desired temperature of HP steam to HCU turbine was about 410°C whereas it came down to level of 350°C. This affected turbine speed with consequences in the process like t'put reduction and production loss. This happened after revamp in HGU in the previous shutdown.
- **Troubleshooting:** All chronological data were studied. It was found that before HGU revamp HP steam temperature to turbine was normal; hence, doubt on sudden deterioration of HP steam line insulation was not raised; however, survey was made through offsite HP steam header; insulation nowhere found broken. HP steam temperature at sources like, power

plant, CCU and HGU were noted; found normal. It became risky to open bleeders of HP steam header in offsite to drain condensate, if any, from chance of uncontrolled opening of the bleeder making this act unsafe like being unable to approach to close the valve when necessary.

HP steam generation flow rate from all sources were noted; it was found that before HGU revamp HP steam ex HGU was mostly adequate to HCU turbine, but after revamp, HP steam generation in HGU came down which was not a setback rather it was an energy efficiency improvement in HGU. But due to this, there was makeup of HP steam from other sources as mentioned and as unlike HGU, other sources as above were at a long distance, the overall temperature of HP steam at HCU turbine inlet came down. This indicated inadequate insulation effectiveness in offsite HP steam header. It was observed that insulation was made of mineral glass wool; as such for such a high temperature, the best insulation material is calcium silicate. It was decided to replace the insulation by calcium silicate. Till the time of completing the replacement, t'put was tried to be managed by increasing steam temperature at power plants to possible extent.

The problem got solved after adequate replacement of insulation as mentioned above.

- **Lesson:** Sometimes the effect of nonstandard job, done once, takes a long time to surface out. Also, there was a gap in design coordination in HGU revamp which overlooked with respect to effect in HCU on revamp of HGU. Hence, multi units' coordination team should be formed and activated in revamp of a single unit.

5.6.7 Leakage in Flare Header in HCU

- **Incident:** In HCU inside the battery limit, flare header got leaked as found when observed that liquid drops falling from the header in one location. The location was inspected by the inspection department and they confirmed that prima facie it appeared due to corrosion in the header. This incident indicated that the unit needed to be shut down but the refinery could be kept running as the leak was within battery limit of HCU; however, shutting down HCO meant quantum loss of diesel production for about two weeks including startup, shutdown and replacement of leaky flare header segment.

- **Troubleshooting:** Due to the above threat of production loss, alternative ideas were being explored to come out of the situation without shutting down the unit. One alternative was hot tapping in the header, but unfortunately, it was observed that there was a branch of flare line joining the header in close vicinity of the leak in the header; hence, hot tapping would not serve the purpose. As the leak was very small like little drops of hydrocarbon along with hydrocarbon vapour, maintenance was asked whether pinning technology of dispersing metal from nearer zones to the place of leak thereby closing the pin hole leak was possible or not. Inspection was asked to inspect the location again to confirm the thickness level in that vicinity was adequate or not to apply the pinning process. Inspection gave hope and maintenance carried out the job and the leak stopped.

  Unit operation thus could be maintained for the time being and replacement of that section in the flare header had been planned to be carried out in next maintenance shutdown, however, sourcing of material and agency for execution had been kept lined up to expedite in exigency, if any.

- **Lesson:** Generally in the section of elbow/reducer, thickness is kept higher than in the straight length of the pipe as the chance of thickness reduction in that section is more. Similarly, for vital utility headers like flare lines, steam lines, cooling water lines where branch tap off is taken, thickness in that zone should be kept higher than straight run length as hot tapping is not executable in that zone.

5.6.8 Medium Pressure (MP) Steam Line Leak in Refinery Common Header

- **Incident:** When such leaks were reported, agencies were explored for hot tapping in the line on line to establish bypass of the leaky zone and isolating the leaky zone by providing an isolation valve. No agency was agreeing to do hot tapping at such a high temperature of about 180°C and pressure of about 14 kg/cm^{2}g. However, one agency agreed and convinced through their credential of capability but no example they could give as to whether they did hot tapping jobs at such high temperature or not. It was a matter of uncertainty but decided to take a trial as technology was convincing. Availability of valves, pipe segment, and flange was checked but so many fittings were not available; however, it was found that leaky zone could be isolated without shutting down the whole refinery but HCU would have to be shut down as MP steam was used in the stripper of HCU. Also, isolation of offsite MP steam lines would lead closing of heating to vacuum residue tanks. This indicated diesel production loss of about 10 days for shutdown, start-up of HCU and execution of the steam leak rectification as mentioned above.

- **Troubleshooting:** It was found that HCU also generated MP steam which was joining the refinery MP steam header outside its battery limit and also the net amount of steam export was just a neck to neck requirement of stripping steam to the stripper. Also, it was found that valve and fittings requirements for hot tapping in that zone were available. Hence, it was decided to carry out hot tapping inside HCU battery limit and then maintain HCU operation by manually maintaining unit MP steam generation to meet the stripping steam requirement. Subsequently, isolate the MP steam line in the offsite to carry out the leak rectification with minimum fittings like just with a spool piece of that dia in that line.

  The first part was carried out successfully and with vigilant monitoring of HCU, manual stripping control in the stripper was also successful. Hence, offsite MP steam header inlet battery limit valve could be closed without anxiety with respect to stability of HCU operation, and the problem got solved.

- **Lesson:** Not only one solution, but innovation to overcome a series of problems sometimes is required to come out of the crisis.

5.6.9 Thickness Reduction in Refinery Natural Gas (NG) Header in Multiple Locations

- **Incident:** When a regular thickness survey on refinery NG header carried out by inspection department, they informed that thickness has reduced considerably in all the elbows of the header and in many places, it came out close to allowable thickness for operation. It became a matter of anxiety that refinery P&U fuel and HGU feed processing would have to be switched over to naphtha which would be a huge loss of profitability to the refinery when so many spare elbows were not available in the stock. Moreover, as per agreement with NG supplier, compensation needed to be provided to supplier against non-consumption of NG.
- **Troubleshooting:** Various alternatives were explored to come out of the crisis. It was found that non-metallic polymers tape came out in the market and applied in the pipeline network in such a situation. But as a protocol of getting it accepted by the refinery inspection department as a part of quality control of the job, when they were discussed with them, they did not agree. Then, an accredited certifying agency like Engineers India Limited (EIL) was approached and they confirmed that their quality control department had not yet regularized that application in their approved list and kept it still on experimental stage.

  Suddenly, it was decided to explore the quality details of all materials including those elbows which were used for execution of these NG lines. When looking at the records, it was found that elbows were selected with those levels of thicknesses which were found during thickness survey, i.e., there was no thickness reduction during course of time. This confirmed that during construction, elbows of thickness same to that of straight pipe were selected instead of using one level higher as it should be as per inspection department.

  However, as it was confirmed that there was no thickness reduction from original level and existing thicknesses were above minimum level though close to minimum allowable, it could be concluded that thickness will continue to remain unchanged. However, it was decided to keep regular monitoring of thickness during operation. This analysis and conclusion helped to continue operation as such and saved loss of profit.

- **Lesson:** Sometimes problem after thorough analysis does not appear to be a real problem.

5.6.10 Oil Film Contamination in Storm Water Drain Moving Out of Refinery Raising Environment Complaint

- **Incident:** Without reason, storm water flow rate in the refinery was higher than normal due to fire water line leakage and adding to storm water drain; this storm water through refinery open channel flowing down to nearby river but passing through drain across the village paddy field. As usual sometimes out of control some oil slipped into this storm water drain from the process units floor showered by heavy rain where peripheral drains of process units could not block the oily water from entering into the storm water drain.

  Though this storm water was meant to flow down to the river, villagers used to consume this water for their paddy field, but when there was oil contamination, they could not use it and they became vigilant to highlight that oil contaminated water from the refinery was polluting the river.

  It became a real threat to continue refinery operation during such a situation.

- **Troubleshooting:** While sensitizing operating crew to prevent ingress of oil into storm water drain, several traditional methods of trapping this oil like providing oil catchers made of jute, husk, etc., but at intermittent locations, but even with this, oil could not be trapped fully. On analysis, it was found water velocity in the storm water drain was high enough restricting oil to skim out and oil catchers to function properly.

  While there was no other alternative than to tackle the situation by strengthening traditional methods for the time being, preparing a storm water pond inside the refinery was decided to allow settling of storm water thus ensuring total catching up of entrained oil in water and allowing oil free water moving out of the refinery.

Later on due to provision of this facility, part of this water from the storm water pond could be recycled to the refinery cooling water system based on its quality and based on the cooling water treatment program in the cooling tower.

- **Lesson:** Troubleshooting of a problem sometimes contributes in getting bonus achievement out of it.

5.6.11 In P&U, Naphtha Pump Suction Strainer Getting Choked Frequently

- **Problem:** For supplying naphtha to GT, light naphtha from HCU routed to offsite tanks under control of P&U. Naphtha from this tank is pumped by a centrifugal pump with one as standby to supply naphtha to GT. Pump suction strainer was getting frequently choked; on checking the dirt in the strainer, it was observed it was black fine dirt iron like particles. The tank was continuously in line and was opened to the atmosphere for maintenance. Frequently choking of strainer caused uncertainty on naphtha pumping to GT with consequent uncertainty in power generation.
- **Troubleshooting:** HCU operation was checked; its naphtha recovery section was not steady, i.e., there was slippage of H_2S from de-ethanizer bottom; the problem was continuing for about 3 to 4 weeks. Major part of this naphtha was routed to excess naphtha tank, another part was blended in MS pool and remaining part was routed to tank for supply to GT as mentioned above. It was observed that specification of naphtha tank for selling naphtha did not fail as a major part in the blend was straight run naphtha. MS tank quality failed in Cu-strip corrosion; then it was doubted that this H_2S contaminated naphtha was internally corroding the tank and taking the corrosion product to pump suction to choke the suction strainer with consequent upset in pumping. The problem got solved after taking direct suction of straight

run light naphtha from the refinery when by that time the contaminated naphtha tank evacuated and flushed out.

- **Lesson:** Connection to multiple sources of naphtha from GT saved the refinery from unwarranted shutdown.

5.6.12 In Offsite, Internal Fuel Oil (IFO) Pumps Getting Vapour Locked Causing Problem in Furnace Fuel Firing

- **Problem:** Fuel oil burners in unit furnaces were getting extinguished intermittently; on checking, it was observed that fuel oil PCV upstream pressure was fluctuating. While talking to the offsite pump operator, the discharge problem in the pump was informed. When visited offsite pumps along with maintenance people, pump discharge pressure fluctuation observed even with standby pumps; even by water pouring on pump casing could not solve the problem. This created uncertainty in IFO supply to furnace burners with consequent threat to refinery t'put.
- **Troubleshooting:** No dirt found in suction strainer; suction fluid temperature was high enough, hence, no reason of high viscosity problem, particularly when it could be known that % of HCU slop cut of low viscosity in the internal fuel oil (IFO) blend was more as compared to residual fuel oil (RFO) ex DCU. It was decided to go to IFO tank. It was seen that IFO tank was so hot that its wall could not be even touched. While exploring the reason, it was found that in HCU as hydrotreated diesel pour point was running high, to improve its pour point, slop cut draw rate increased; but slop cut run down cooler capacity was limiting which resulted in very high run down temperature of slop cut. HCU t'put partly reduced; from DCU along with RFO, heavy coker gas oil (HCGO) added and routed to part empty standby IFO tank and the tank made ready quickly, and problem got solved.
- **Lesson:** It was due to lack of monitoring of IFO tanks in the field and its temperature from the tank monitoring control room.

5.6.13 In HCU, Inert Gas Blanketing Pressure Came Down

- **Problem:** Inert gas blanketing was very important in HCU and being used in a number of vessels to control in and out of liquid in the vessels. But one day, its pressure came down causing difficulty in controlling operation in those vessels; moreover, alternate fuel gas makeup facility was not provided.
- **Troubleshooting:** When P&U was contacted to increase N_2 gas generation, they expressed inability because their N_2 evaporator capacity was limiting; they also told that gas N_2 generation was normal and doubted that there must be some loss in the unit. No loss source inside the unit could be detected; when enquired about the layout of this inert gas line, it came out that a part of this inert gas was routed to coker distillate tanks which were of fixed roof design with inert gas blanketing facility. Tank roof top visited and it was observed that there was continuous bleeding out of vapour from the tank breather valve and inert gas PCV was open; actually, PCV was not working properly resulting in continuous flow of inert gas to tank blanketing with consequent venting out through tank breather valve. The problem got solved after rectification of PCV.
- **Lesson:** Tank PCV indication brought to the tank monitoring control room; also, fuel gas blanketing facility was provided as alternative to ensure HCU operation in absence of N_2 plant availability.

5.6.14 In Offsite, Medium Pressure (MP) and High Pressure (HP) Steam Lines Condensate Draining and Joint Leak Rectification Causing Refinery Shutdown

- **Problem:** In one case, when offsite low point drain (LPD) valves opened for condensate draining before MP or HP steam lines commissioning, these bleeder valves were not approachable to make these closed; thus there was huge noise in the refinery and it was difficult to operate refinery with such noise.

In another case with a refinery in normal operation, one LPD in the MP steam line got leaked and there was huge noise causing difficulty as mentioned above.

- **Troubleshooting:** In the first case, instead of further approaching startup activities, steam header valves got pinched opened; these bleeder valves closed, then standard hose connection made to these bleeders and steam header valves opened again. Thus condensate draining could be controlled.

  In the second case, it was decided that instead of conventional butt-welding, a reinforcement pad should be provided while connecting a LPD in the line and also the area where LPD is located, grass should be removed and area should be concreted. The problem was eliminated with these improved practices in future.

5.7 Miscellaneous Equipment and Fittings

5.7.1 For PSV Isolation, Twin Valve or Separate Block Valves Provided Got Stuck Up

- **Incident:** Twin valve common wheels could not be rotated to switch over to stand by PSV; in the separated isolation valves case, block valve gate found stuck up.
- **Recommendation:** Nowadays, twin valves are not provided and the gate valve should be placed with its stem and wheel in vertically downwards.

5.7.2 PSV is Passing Causing Continuous Loss through Flaring

- **Incident:** This problem is observed in many process units starting from units handling gas from column/vessel with clean fluid up to units where viscous/congealing liquids are in operation. In a vessel, generally a demister pad is provided at the top to avoid entrainment in the vapour to ensure not only liquid carry over to the downstream process but also to protect PSV from passing problems. Even then, as mentioned above, PSV passing problem is observed.

- **Recommendation:** Standby PSV should be in place with an isolation facility to each PSV. To ensure easy maintainability, now a day, the main PSV is provided along with pilot PSV whose pop setting is kept slightly lower than main PSV so that main PSV popping is not only avoided but also in case pilot PSV could not be reset after popping only it can be taken to workshop on which maintenance job becomes simpler.

  For units handling with viscous/congealing liquid, there should be one rupture disc between PSV and its isolation valve with one pressure transmitter in between PSV and Rupture disc so that in case of pressurization rupture disc will fail first to avoid popping of PSV because if PSV pops there is a chance of continuous passing without getting reset due to possibility of entrained liquid getting deposited inside PSV. Whenever there is any pressure alarm in the control room from the pressure transmitter in between PSV and Rupture disc, it indicates that the rupture disc has failed, and that time its PSV is to be isolated and the rupture disc should be replaced.

5.7.3 In Hydrotreating Unit, Compressor Suction Valve When Tried to be Opened, Huge Gland Leak Started

- **Incident:** Had to be handed to maintenance and the stand-by was to be taken in line. Hence, the suction valve of the stand by compressor was to be opened. When tried, a huge gland leak started and the operator immediately closed the valve; with this situation, the unit appeared to be brought to shutdown.
- **Troubleshooting:** Valves have two gland seals in stuffing box packing. Hence, though after part opening a huge gland leak was observed, still quickly, the valve made full open when gland leak stopped which indicated that primary seal was working and thus unit operation could be continued.
- **Lesson:** Gate valves should never be kept in partial open condition; with this habit, secondary seal life gets reduced. It should always be kept fully open. However, in this case, the secondary seal got damaged, a nitrogen should be kept plying

on the valve and the valve should be replaced at earliest opportunity.

5.7.4 Frequent Tripping of Critical Pumps/Compressors Causing Unit Shutdown

- **Incident:** This has happened by respective motor tripping where neither any process causes nor any fault in the motor with its protection sensors in the field.
- **Troubleshooting:** Root cause identified in electrical substation like relay or circuit breaker. Regarding relay, it may be mentioned that from a reliability point of view electronic relay found better than earlier installation with electro-magnetic relay. Sometimes failure of the circuit breaker also caused the tripping which needed to be replaced by a new one.
- **Lesson:** In troubleshooting, all disciplines of maintenance should be engaged in discussion to let them understand whether there can be any chance of problem on their side; this helps to reduce time in problem identification thereby saving production loss.

5.7.5 ID Fan/FD Fan Tripping Causing Unit Shutdown

5.7.5.1 Case I

- **Incident:** In some cases two FD fans are provided and it is observed that one is in operation and other kept as standby to save energy consumption or as per maintenance request.
- **Recommendation:** It should be an operation philosophy of running both at 50% capacity so that in case of one tripping other can take 100% load automatically as per control system.

5.7.5.2 Case II

- **Incident:** In some cases, FD fan without spare is provided; however, an ID fan is always found to be provided without spare. Generally with variable drive in it, scoop oil systems in motor-fan coupling sometimes malfunction thus preventing to increase IGV (internal guide vanes) opening or to control fan speed when necessary; sometimes, motor amperage becomes critical and sometimes vibration problem. All these were normal once in operation but sometimes it trips on start-up high current.

- **Recommendation:** As for the process initial load requirement is low, a soft starter switch instead of normal start-up switch should be provided to start the motor to avoid overload tripping.
  However, a motor with variable drive is also preferred.

5.7.6 Cooling Water Valve Gate Detached and in Other Case Flare Header Valve Gate Detached in One Unit

- **Incident:** These were two different incidents that happened during the unit startup after M&I shutdown. In both the cases units could not be operated due to non-availability of cooling water flow and non-thoroughness of the flare system respectively. Maintenance departments were called to rectify on line as the cooling tower system was already commissioned and stopping the cooling water system meant shutdown of other running units. Similarly, for rectifying battery limit flare isolation valve of unit, total flare header could not be decommissioned as other units were in operation; so maintenance was asked to rectify the flare valve on line. But they could not, and the refinery came under shutdown threat. But that was a critical time with respect to supply chain management and the refinery had to suffer heavy losses.
- **Troubleshooting:** Cooling water valves are also exposed to dissolved oxygen in it thus with continuous contact of water and air, rust occurs in hook up location between valve sit and stem thus sit may get detached sometimes; though same logic is not applicable for flare valve, it might happen due to H_2S corrosion or might the valve be stored in bad ambience before installed in position. The situation for the time being was managed by hot tapping for providing a new valve by passing the stuck up valve thus shutdown of running units could be avoided.

Based on the above analysis, it was decided to install a valve with its wheel/stem in downward direction for all cooling water valves and flare valves so that even in case of sit detachment from steam, if any, the seat will move downward by gravity thus not blocking the flow path.

- **Lesson:** Unless the above installation guidelines are religiously followed, emergency hot tapping causes loss to the refinery as the job agency mobilization and execution takes 7 to 14 days, which is costlier to refineries.

Chapter 6

Troubleshooting Case Studies in Technologies Operation

A description of the processes of various technologies has been avoided assuming that the readers and engineers are well aware of the same as a prerequisite to proceed for troubleshooting of the problems in operation of the concerned technologies. Troubleshooting in various technology processes was another kind of experience in dealing with any problem/trouble. Sometimes a lot of probable causes were found, and listing all case studies restricted to each single cause will make the documentation a quite bulky. Hence, instead of listing out all case studies, effort has been made to demonstrate the troubleshooting in process units through listing of all troubles/problems with all probable causes as given below. The operating crew would have to analyse, eliminate and arrive at particular cause/causes. Almost all technologies have been covered in this book with respect to troubleshooting of the problems as follows:

6.1 Fluid Catalytic Cracking Unit (FCCU)

Due to the process with the circulating catalyst system along with continuous regeneration facility, it has become one of the most complicated units, as troubleshooting has become a regular habit of

Petroleum Refineries: A Troubleshooting Guide
Dhananjoy Ghosh

ISBN 978-981-4877-81-7 (Hardcover), 978-1-003-18040-1 (eBook)
www.jennystanford.com

unit operation. Hence, it should be wise to highlight the methodology of troubleshooting in FCCU operation as mentioned below. In FCCU, troubleshooting capability is the most important criterion to operate the unit successfully; the troubles are as frequent as observed and in such varieties in nature that it is necessary to discuss troubleshooting methodology separately for this unit before discussing the problems encountered. Also, instead of discussing case studies, probable causes of each problem are discussed as several times multiple reasons found to be the causes of the problem.

Specific troubleshooting methodology in fluid catalytic cracking

- ✓ Understanding of Process design parameter.
- ✓ Understanding of design details of each and every component of reactor-regenerator.
- ✓ Establish steady state test run data which may not be limited to only following areas rather to be made more exhaustive :
    - Product yields and composition
    - Detail performance data of each equipment
    - Flow/temperature/pressure profile across major rotating equipment, catalyst circulation system, stand pipe, cyclone, utility system around reactor and regenerator, reactor vapour line, main fractionator bottom circuit, across slide valves, regenerator flue gas circuit, major distributor, feed nozzles, air grid
    - Test run cycle/survey of equipment performance – based on experience
- ✓ Ensure installation of all components as per licensors approved final drawing during shutdown.
- ✓ Avoid using improper refractory and refractory lining and any patch repairs may lead to repetitive failure.
- ✓ Ensure 100% leak test around Rx-Rg.
- ✓ Avoid using improper insulation around the reactor and vapour line causing rainwater ingress creating a cold spot.
- ✓ The fluid catalytic cracking (FCC) process is very complex due to continuous catalyst circulation and its interactions with hardware, feed, and products.
- ✓ Reverse the changes in operation to what have been followed previously i.e.; just before the problem.

✓ It is not difficult to determine the cause of a problem with systematic approach.
✓ Troubleshooting is done by gathering data on current operation and comparing this information with data from test run/equipment survey data/recent historical data.

6.1.1 Case Studies/Problems due to Different Probable Causes

6.1.1.1 Poor reactor performance

- Catalyst contamination with metals
- Poor mixing of catalyst and feed in the riser
- High delta coke on regen catalyst
- Inadequate feed dispersion steam
- Malfunctioning of feed distributors
- More unstripped hydrocarbon with spent catalyst
- Lower catalyst make-up rate
- Change in feed mix
- Change in feed characteristics
- Change in fresh catalyst formulation and micro-activity test (MAT).

6.1.1.2 Poor regenerator performance

- High coke level in regen catalyst
- Very high throughput
- Heavy feed
- Inadequate oxygen/air
- Improper dense bed level in regenerator
- Higher regenerator temperature or increased after burn due to poor stripping of spent catalyst
- Damaged catalyst entry system
- Damaged air entry/distribution system

6.1.1.3 Poor cyclone performance and high catalyst loss

- Very High feed rate
- Higher rate of fines generation during catalyst circulation

- Catalyst level is not proper
- Internal refractory or metallurgical component damage
- Excessive catalyst loading in cyclones
- High inlet or outlet velocities
- Coke or refractory plugged dip legs and immovable dip leg outlet flappers
- Soft fresh catalyst or catalyst with an abnormal particle size distribution
- Eroded barrels, hopper cones and dip legs, weld crack in cyclone

6.1.1.4 Indication of poor regenerator cyclone performance

- Increased fresh or equilibrium catalyst loadings to maintain unit inventory
- Decreasing regen level at normal catalyst loading and at constant reactor level
- Increased fines catch from the electrostatic precipitator
- Need for more frequent soot blowing to maintain heat transfer in the waste heat boiler or CO boiler
- Increased flue gas stack opacity
- Shift in the equilibrium catalyst analysis 0–40 micron, 0–80 micron or average particle size distributions
- Increase in MAT and decrease in metal contaminants due to higher fresh catalyst loading

6.1.1.5 Indication of poor reactor cyclone performance

- Increase in main fractionator bottom pump-around or slurry product streams' basic sediment and water (BS&W) or ash analyses.
- Decrease in heat transfer in main fractionator bottom pump-around exchangers. (Catalyst settling or fouling causing increasing pump around return temperature at constant flow.)
- Decrease in slurry product rundown pumping capacity due to catalyst settling in piping and exchangers.

6.1.1.6 Flow reversal in reactor-regenerator

To understand the problem, let us show the control scheme of reactor-regenerator loop as follows:

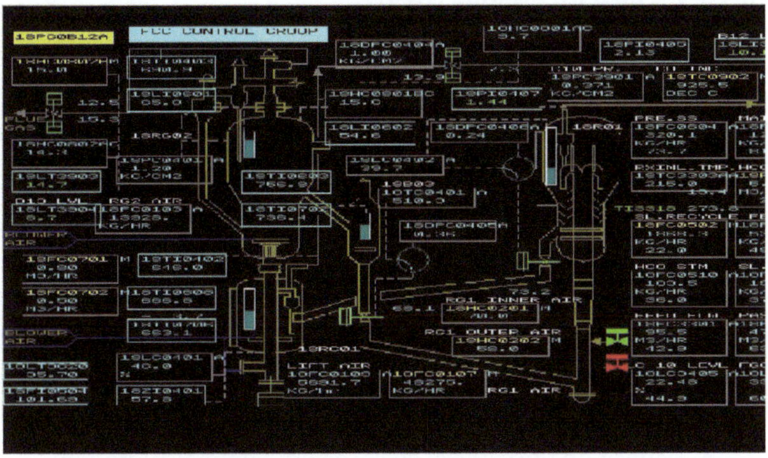

In the above diagram, the left side equipment is a two-stage regenerator with the first stage at the bottom and the second stage at the top (RG). On the right side, it is a reactor (RX). The green coloured valve in the downward catalyst transfer line from reactor to regenerator is called stripper catalyst sliding valve (SCSV) and the green coloured valve in the downward catalyst transfer line from second stage regenerator's catalyst discharge pot to reactor is called re-generator sliding valve (RGSV). There are four level controllers, viz., 3 in RG and 1 in RX stripper section shown above.

- It is essential to maintain stable differential (dP) pressure across SCSV and RGSV so that catalyst flow can be from RG to RX and from reactor-stripper back to regenerator.
- Negative dP (differential pressure) would cause backflow, causing uncontrolled burning in RG resulting in damage to RG internal, hot flue gas back to RX and MFC (main fractionator) and causing mechanical damage.
- Cause of flow reversal are
  - ✓ loss of air blower;
  - ✓ presence of water in feed;
  - ✓ high catalyst circulation rate;

- ✓ loss of RG or Stripper bed level;
- ✓ failure of RX temperature controller and RX stripper level controller.

6.1.1.7 Reactor cyclone coking

- Coke formation is usually due to condensation of heavy hydrocarbons.
- Material condenses in the 'dead' area behind the cyclone inlet horn.
- Can be minimized with:
  - ➢ good feed injection;
  - ➢ increased steam in riser;
  - ➢ making sure unit is hot when feed is initially injected;
  - ➢ taking feed out before riser outlet temperature is too low.

6.1.1.8 Nickel accumulation in equilibrium catalyst results in

- Deposit on the catalyst surface with minimal migration;
- Catalysis of undesirable dehydrogenation reactions and coke;
- Slight decrease in equilibrium MAT;
- Increase in delta coke and decrease in catalyst circulation rate and cat-to-oil ratio, leading to lower conversion, lower gasoline yield, lower gasoline selectivity, and lower total liquid yield; fuel gas yield will be increased.

6.1.1.9 Vanadium accumulation in equilibrium catalyst results in

- Migrates to the active zeolite components and destroys them
- Decreases equilibrium MAT
- Slightly catalyses undesirable dehydrogenation reactions and coke
- Decreased MAT leads to lower gasoline and LPG liquid yields, lower gasoline selectivity, and lower total liquid yield

6.1.1.10 Catalyst circulation problem in stand pipe

- Standpipe pressure profile is not uniform
- Inadequacy of steam, air, or some other gas being used for aeration

- Steam/air/gas is not absolutely dry
- Could some of the aeration taps be plugged

6.1.1.11 High catalyst loss from regenerator

- Regenerator dense bed level indication failure
- High catalyst density
- Flapper valve malfunctioning
- Erosion at Top of dip-leg by a most common cyclone failure

6.1.1.12 High catalyst loss from reactor

- Regenerator dense bed level indication failure
- High catalyst density
- Flapper valve malfunctioning
- Erosion at Top of dip-leg by a most common cyclone failure

6.1.1.13 Failure of expansion joints

- The design of these expansion joints is not proven
- Improper installation
- Difference between design and actual data
- Corrosion during operation or stress corrosion cracking during shutdown.
- Water or condensate carryover along with purge fluid
- Erosion of lining
- Over the period material reached fatigue life

6.1.1.14 Other miscellaneous problems in FCCU

- Coke/polymerization in the downstream of RX-RG section, i.e., in the debutanizer bottom reboiler due to improper cleaning of gas-con section from rust/corrosion products.
- Stress corrosion cracking in main column overhead condenser due to inadequate washing before opening during shutdown.
- Coke up in the bottom circuit due to delayed decision of shutdown causing stoppage of flow with high catalyst content.
- Deterioration of fresh catalyst performance over the period due to reasons best known to catalyst vendors.
- Corrosion in the main fractionator's top section internals due to condensate trapping in the locations.

6.1.1.15 Poor product yield performance

- Observation: High H_2 yield, high coke yield, higher riser velocities, overloaded gas compressor limiting.
  - Causes: High metals in feed, catalyst contamination, poor reside catalyst, feed type change, low catalyst replacement, vac. unit operation problem.
- Observation: High coke yield and poor weight balance.
  - Causes: Leak in exchanger trains, heavy hydrocarbon in feed or start up valve partly open.
- Observation: Low coke yield, poor weight balance, unsteady feed heater pressure.
  - Causes: Leak in exchanger train or light hydrocarbon in feed.
- Observation: High coke yield after burning in regenerator.
  - Causes: Poor stripping or poor feed vaporization.
- Observation: Low catalyst activity.
  - Causes: High regeneration, temperatures, localized high temperature, high Ni and V on catalyst, excessive steam in regenerator or torch oil.

6.1.1.16 Common failure in cyclone area

- Weld crack.
- Crossover duct crack.
- Vortex termination.
- Catalyst/vapour entrainment due to problem with flapper valves.

6.1.2 Miscellaneous Process Problems and Solution in FCCU

- Improper aeration—too high or too low, aeration point choked may cause low or erratic differential pressure across the slide valves. To overcome, adjust aeration RO valve with previous nitrogen and airflow. Also, aeration point to be dechoked.
- Catalyst is too coarse and fine content is too low. Here, recycling of catalyst fines and slurry to be done; catalyst of different particle size to be added. Also, iron content of the catalyst should be checked.

- Foreign object restricted flow of catalyst in stand pipe. Here, use gamma ray scanning to confirm plugging, and then use steam or nitrogen pressure to dislodge the foreign material.
- Catalyst is defluidized due to its high bulk density or low regeneration pressure. To overcome, increase regenerator pressure.
- Wet aeration air or steam. Here, ensure it is dry to avoid consequences.

6.1.3 Incident Case Studies in FCC Troubleshooting

- **Problem:** Limitation in FCC t'put increased due to limitation of airflow in regenerator.
- **Incident:** FCC t'put could not be increased on account of orifice chamber chokage. Whenever efforts were made to increase t'put, this led to increase in regenerator pressure.
- **Troubleshooting:** It was observed that at t'put of 190 m^3/h regenerator pressure varied between 2.75 to 2.85 Kg/cm^2g. Normal pressure of regenerator is 2.6 Kg/cm^2g. Increase in pressure was due to orifice chamber chokage. Increase in pressure led to decrease in delta pressure across RX-RG. This primarily restricted t'put increase.

  It was thought that an increase in COT to maintain the same reactor temperature would lead to reduction in catalyst circulation leading to reduction in air requirement to regenerator. This decrease in regenerator air would be made up by increasing t'put. Accordingly, a trial run was conducted and found successful.
- **Lesson:** Sometimes alternative solutions should be explored when planning is not favouring to address root cause.

6.2 Delayed Coker Unit (DCU)

Here though there is no catalyst because of it being a thermal cracking process, but lack of effective emergencies handling/process troubleshooting may lead to heavy penalty in refinery operation, as here feedstock is very heavy like vacuum residue (VR) which on mal-operation may coke up furnace thereby reflecting huge long time repair/replacement job. Moreover, VR being heavy feedstock poses threat on unwarranted fire from by chance leak sources because of very low auto ignition point of VR and its heavier conversion products.

Before proceeding to troubleshooting, let us understand the kinetics of coker unit operation as follows:

$$K = (1/t) \times 2.303 \log (100/(100 - X)), \tag{6.1a}$$

$$t = V/Q, \tag{6.1b}$$

where K is reaction constant; t is the residence time for conversion; X is the conversion, i.e., distillate yield in % wt; V is hold up volume of furnace coil or coker drum; Q is the volumetric flow rate of unit feed.

K is dependent on operating temperature and values are available in the literature for different temperatures for thermal cracking. Here, conversion/cracking takes place both in furnace and in coker drum; in furnace, conversion is negligible like 5 to 6% and balance conversion which is major, i.e., at about 65 to 70% occurs at coker drum. So the above formula should be applied twice, viz., first for furnace coil, next with outlet vapour flow from furnace, the equation again to be applied for coker drum.

So, from the above equations, one can find out the value of theoretical conversions and compare the same with actual conversion to recognize the problem in distillate yield, if any.

To demonstrate all troubleshooting in DCU in capsule form, effort has been made to present different problems and possible causes as given below:

6.2.1 Velan Valve on Coke Drum (Under Vapour-Heating) Got Stuck-Up at 60% Opening

- Measures taken to avoid shut-down:
  - ➢ Reduce secondary flow through furnace coil corresponding to heater turn-down ratio
  - ➢ Adjust tubulizing water to heater coil accordingly
  - ➢ Reduce COT by 2–3°C to minimize gas make
  - ➢ Increase recycle ratio by minimizing primary flow; this will also help to reduce gas make

- Adjust column withdrawal accordingly
- Do the change-over slowly keeping watch on system pressure
- Attend the valve in next idle hours of the coke drum

- **Worst scenario:** If reactor pressure tends to shoot up after completion of changeover even after above changes, run the unit for at least 12 h with the crack open of the chamber blow-down valve.

6.2.2 Quench Tower Coking

- Measures to be taken to avoid shutdown:
  - Reduce t'put in proportion to the extent of delay up to heater turn-down ratio
  - Increase recycle ratio by reducing primary flow
  - Reduce COT by 2–3°C
  - Adjust column withdrawal accordingly
  - Start silicon (antifoam) injection much earlier
  - Increase gas oil quench to Coke Chamber outlet line by 10%–20%; this will help to wash the cracked vapour and arrest the carried over coke fines (if any) to quench tower
  - Maintain system pressure slightly on higher side than normal during change-over; this will help reducing vapour velocity temporarily through coke drum thereby arresting entrainment of coke fines because of high coke level
  - Slowly do the change-over to avoid pressure surge

6.2.3 Heater Skin Temperature Running High but Avoiding Shutdown to be Tried

- Measures to be taken:
  - Reduce COT by 2–3°C (even more for the affected pass)
  - Watch the combined COT at the inlet of coke chamber (>480°C)
  - Reduce pass flow (but not below turn down flow)
  - Change over gas burner to oil burner to the affected areas or take out the burner from the affected area

- Increase excess air if it permits
- Increase column bottom temperature to the extent possible by adjusting the temperature C/V
- Regular monitoring of skin temperature by temperature gun
- Visual inspection of heater tubes for any hot spot frequently

6.2.4 Heater Inlet Pressure Running High and Try to Avoid Unit Shutdown

- Measures to be taken:
  - Reduce COT by 2–3°C (even more if required)
  - Watch the combined COT at the inlet of coke chamber (should be >480°C)
  - Reduce pass flow (if possible)
  - Inspect heater frequently for any hot spot
  - Increase column bottom temperature to the extent possible by adjusting the temperature C/V
  - Regular monitoring of skin temperature by temperature gun
  - Increase boiler feed water (BFW) flow slightly to the affected pass
  - If feed viscosity high, reduce if possible and/or operate with marginal increase in coke drum level with reduction in furnace COT

6.2.5 Coke Drum Bulging/Cracking

- Coke drums tend to suffer from bulging phenomena due to the cyclic nature of their service. Water quenching plays an important role in coke drum bulging. High quenching rate, with thermal gradient over 10° F per inch, results in more bulging. To measure this effect, a unit quenching factor (UCF) is developed, which is the ratio of water quenching time in minutes to the coke yield per drum in MT. when UCF is greater than 0.5, bulging is minimum. It is better to maintain UCF at 0.8 to achieve a longer life of coke drum.

6.2.6 Coke Bulk Density is Low

- Probable causes are mostly due to feed becoming lighter reflected by low CCR/density/viscosity of feed. Sometimes slop processing may result in low bulk density of coke.

6.2.7 Coke Yield is High

- Main reason is the higher recycling ratio. Recycle ratio should be as low as possible but should not be below design value. Generally operating crew keep it in the range of 0.15 to 0.20 from fear of furnace tube coking and reduced cycle length. But at this ratio, it is difficult to get coke yield below 30%. Ideally coke yield should be as low as 25%. However, ratio should be compared with design ratio; even if with design ratio, coke yield is high, then alternative approaches to be followed like increasing furnace COT or increasing coke drum level as much as permissible.

6.2.8 Unit Cycle Length is Low

- Probable causes:
  - Some skin thermocouples malfunctioning.
  - Higher reading of one or two skin thermocouples ignored.
  - Furnace tubulizing water flow meter malfunctioning and actually flow is lower than minimum desired.
  - Furnace outlet temperature necessarily kept higher to get desired distillate yield instead of addressing feed quality. It may be noted that if vacuum column tray efficiencies are good enough to get better separation to achieve low yield with heavier quality of vacuum residue (VR), then by dropping distillate to VR to reduce its viscosity does not help in reducing furnace coil delta pressure and results in reduced run length of furnace, as after VR is formed, it is not easily miscible with distillate. Thus distillate causes higher pressure due to vaporization and heavy VR as usual cause higher thermal cracking. In that case, increase coke drum level as much as permissible.

6.3 Visbreaker Unit with or without Soaker Drum

Like DCU, a visbreaker unit (VBU) is also a thermal cracking unit except that here residence time for partial cracking is limited only to furnace with achievable distillate yield of 6% to 8% max and residue remains at liquid stage but with reduced viscosity due to olefin bonds formed in the residue. Where soaker drum is provided, partial cracking beyond furnace coils is allowed to occur in a drum called soaker drum giving an additional residence time of 20 to 30 min to increase severity of thermal cracking to get more distillate yield up to 12% to 15% and thus with less residue yield but still in liquid stage with lower viscosity due to reason as mentioned above. Hence, it is a bit simple process unlike DCU, but reaction kinetics would remain the same as shown in the DCU section. Due to operation at high temperature like DCU, upset like unwanted coking and residue, VB Tar quality problem may occur, and also due to higher feed viscosity like DCU, there is a chance of congealing in many sections with consequent choking and upset in operation remains.

Some salient troubleshooting experiences in this unit are shared as follows:

6.3.1 VB Tar Drop Test Fails

- **Incident:** VB Tar is a main constituent in fuel oil blend. Sometimes, it is found that one important quality test of fuel oil, i.e., drop test, fails.
- **Troubleshooting:** It is the quality of VB Tar which caused failure of this test. In fuel oil, asphaltene should be well dispersed in aromatics; drop test would fail when asphaltene precipitate instead of dispersing, and this happens when aromatic concentration in VBU feed is less or furnace severity is higher.
- **Lesson:** In this situation, sometimes reducing furnace outlet temperature solves the problem. In some cases even concentration of aromatic had to be increased by adding more

extract in VBU feed blend from vacuum distillate's aromatic extraction unit.

6.3.2 Furnace Coil Delta Pressure Increase Reducing Cycle Length

- Recommendations:
  - During such situations, coil tubulizing water flow meter should be checked whether flow is less or not. Even if it is okay, flow should be tried to increase if scope exists.
  - If it is found that distillate yield is higher than expected, then coil outlet pressure should be reduced by 1 to 1.5 kg/cm^2 by changing the setting of pressure control valve at furnace outlet because lower pressure increases velocity in furnace coil, i.e., reduces residence time in furnace coil thereby reducing thermal cracking inside coil with consequent enhancement of cycle length.

6.3.3 Strainer of Fractionator Bottom Pump with VB Tar Service Gets Choked Frequently

- **Recommendation:** This is due to higher severity of cracking or due to heavier feedstock; it should be tackled by the way as explained earlier.

6.3.4 Line Congealing

- **Recommendation:** This happens due to inadequacy of steam tracing in the pipeline caused by steam trap malfunctioning, particularly in the offsite area. Inside the unit, there is normally more watch on steam tracing effectiveness due to operators being dedicated for a unit of small area. But in offsite, though the area is vast but due to less frequency of job calls, generally operators' watch is found missing. Hence, though steam tracing is a bit economical as compared to electrical heat tracing, still for offsite pipeline, electrical heat tracing should be followed due to reason as mentioned above.

6.3.5 Fire Taking Place during Start-Up or during Operation

- **Recommendation:** Start up being an unsteady state operation, there is every possibility of thermal shock in equipment and due to this, channel head flanges of exchangers with VBU feed and VB Tar catches fire. Hence, steam ring should be provided in these exchangers which should be kept on before start-up and by the time operation gets stabilized, the small amount hydrocarbon leaks, if any, gets coked up in these flanges and smoke found to get disappeared when steam to steam ring can be stopped to keep area clean.

6.3.6 Fractionator Gets Pressurized in Top Section

- **Recommendation:** This is a typical phenomenon in VBU fractionator as observed in a refinery where fractionator tray flooding occurred near the top zone, i.e., at about 25th tray out of 37 trays it was observed because, due to characteristic operating conditions in VBU fractionator, water gets accumulated in that tray and hence, generally water drawing facility is provided where water is collected to a vessel outside from where it is drained out under level control of that vessel. This water shot in the top zone sometimes caused tray dislodging in the top section. Operating crew many a time become lackadaisical to monitor this vessel level till it is critical; the vessel is generally located at technological platforms, i.e., at high elevation, that's why they become indifferent as this problem comes once in a blue moon.

6.4 Hydrocracker Unit (HCU)

Hydrocracking is the most severe hydrotreatment process so far commercially proven and is in use since the last three decades. Here, it is a fixed-bed catalyst system unlike FCC and hence in that direction ones come across with less trouble, but as hydrogen at very high pressure of about 180 bar is involved, lots of safety interlocks are provided in this system and this causes sometimes spurious tripping also. Let us first review hydrocracking kinetics as described below before approaching for troubleshooting.

Hydrocracking feedstocks are heavy atmospheric gas oil, vacuum gas oil, deasphalted oil (DAO) and coker gas oil. Hence, the molecular composition of hydrocracker feedstock becomes mixture of long chain paraffin up to about C20 carbon atoms with or without sulphur and nitrogen, olefins (if coker gas oil is present in the feed), naphthene with side chains with or without sulphur and nitrogen in the side chains, thiophene with side chains with or without sulphur and nitrogen, pyridine with side chains with or without sulphur and nitrogen, aromatic with side chains (paraffinic or olefinic) with or without sulphur and nitrogen, olefin chain with or without sulphur and nitrogen.

Regarding kinetics, hydrocracking is called HDS + HDT + HCU, i.e., hydrocracking reactions are mostly in series with some parallel reactions as shown below:

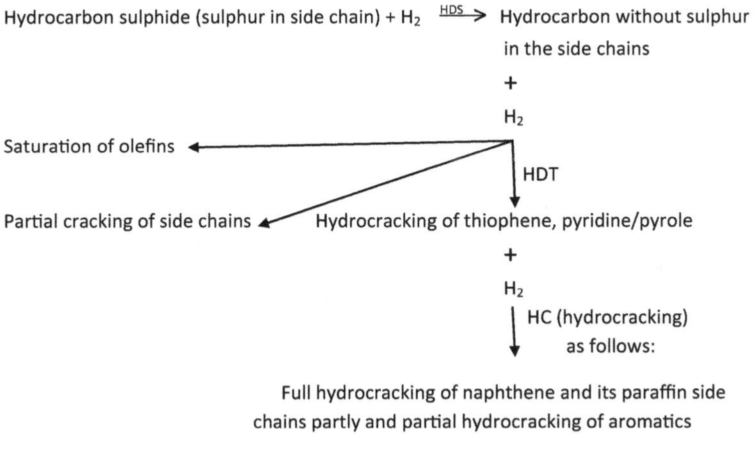

Note:

(1) HDS, olefin saturation, HDT, HC follow first order kinetics; only denitrification follows second order kinetics.

(2) For hydrocracking of heavy gas oil feed mix as mentioned above, operating pressure, temperature, Hydrogen purity and H_2/HC ratio in hydrocracking are about 180 Bar, 390°C, 96% and 300 kg.moles/(m^3/h), respectively. Approx. H_2 consumption: 3% wt of feed rate.

Typical troubleshooting experience in HCU is of paramount importance as it contributes the highest share of profitability in the refinery. A glimpse of troubleshooting in HCU is given as follows.

Figure 6.1 Hydrocracking reactions fundamentals.

6.4.1 Second Stage Reactor's Top Bed Delta Pressure Rising Faster during Course of Operation:

- **Problem:** Even after skimming the top bed in second stage, its delta pressure rose faster though feed rate/quality normal; no feed filter in second stage unlike first stage may form coke in column bottom by repeated power failure leading coke to second stage feed.
- **Solution:** Cartridge filter installed at second stage feed pump suction, and problem got solved.
- **Lesson:** Filter required for each stage feed.

6.4.2 PSA Valves in Hydrogen Unit Getting Damaged/ Replaced Frequently

- **Problem:** The above incident led to PSA tripping.
- **Solution:** Installing a knock out vessel before PSA solved the problem.
- **Lesson:** Always do hazard operability studies (HAZOP) study of each piping and instrumentation diagram (P&ID).

6.4.3 Gas, Naphtha and LPG Make Increase with Decrease of Middle Distillate Yield

- **Problem:** With normal severity and feed rate in both stages, it was seen that gas, LPG, and naphtha increased with less kerosene/diesel yield. It was observed that feed was lighter than design; in the second stage, mainly cracking catalyst and in first stage mainly treating catalyst with less cracking catalyst was provided. Lower severity in the first stage caused less conversion in the first stage with more conversion in the second stage resulting in more Gas, LPG, and Naphtha make.
- **Solution:** Increasing severity in the first stage, decreasing severity in the second stage solved it.
- **Lesson:** Change catalyst formulation as per feed quality to make both stages cycle length same.

6.4.4 Rundown Gas Oil Pour Point was High at –3°C against Normal ppt of –6 to –9°C

- **Problem:** Generally gas oil from HCU gives pour point of –6 to –9°C facilitating its blending with higher pour point straight run gas oil in the tank. But this time it became a problem as mentioned above causing threat to reduction in refinery diesel production.
- **Troubleshooting:** To control HCU gas oil ppt, a bleed called slop cut was drawn from the fractionators. It was found that to get the gas oil pour point, slop cut draw rate had to be increased which meant that conversion in the second stage reactor had come down. As the feed was lighter, increasing severity in second stage, lighter cuts generation was increasing; as such second stage activity was normal; it was difficult to get selectivity with such lighter feedstock; hydrogen recycle ratio was increased but no significant change on the contrary it was reducing second stage t'put. It was decided to make feed quality heavy from VDU and for the time being gas oil pour point was maintained by drawing more slop cut thus sacrificing minor diesel yield instead of jeopardizing the whole tank quality.

6.4.5 Low Cetane Number (CN) in HSD

- **Problem:** While cetane index (CI) just met the spec., CN was failing; by reducing high speed diesel oil (HSD) yield, CI increased which resulted in delta rise in CN; by increasing severity also, CI and CN marginally increased but more hit in profitability.
- **Solution:** As CI was meeting spec., cetane improver dosing started which led to a limited delta rise in CN for fixed CI to meet the spec.
- **Lesson:** Cetane improver inventory should be kept at normal level.

6.4.6 First Stage Reactor Catalyst and Internal Got Damaged Leading Uncertainty in Unit Availability

- **Problem:** Unit was being operated with higher delta pressure in the first stage reactor; though overall delta pressure was within allowable limit; first bed delta pressure was marginally higher than allowable limit while second and third bed delta pressure were within allowable limit. The licensor was informed. They suggested to survey first bed delta pressure with standard pressure gauge, but fixing pressure gauge by opening the bleeder cap appeared to be risky. It was decided to continue operation if that delta pressure would hold at that level as a spare catalyst was not available in stock. After one emergency power failure shutdown, when unit was restarted, first stage delta pressure further rose and feed flow came down due to back pressure; feed withdrawn; gas circulation started; even with gas circulation, delta pressure was unexpected high; on next trial of feed cut in, very low feed rate observed with very high delta pressure; hence, there was no alternative than to shut down the unit and open the reactor. When opened, it was found that the first stage catalyst got lumped to a large extent and also the hydrogen distributor tray in between the first and second bed got cracked partly in the periphery.
- **Troubleshooting:** Refinery was in crisis. Licensor was consulted. The second stage spent catalyst was available in store. As per licensor analysis, spent catalyst of second stage should be able to function if used in the first stage as an ad hoc measure till procurement for first stage catalyst is done. But regarding the repair of the internal, the Maintenance Department informed that welding could not be done for the crack in the internal as mentioned above as a part of horizontal plate of the tray had distorted down which could not be straightened due to space constraint and instead they could provide welding to fix a weir like plate where tray bottom was welded but the vertical plate top could not be welded but it was raised to a height to act as a dam to prevent short circuit of hydrogen through the clearance. Job done, but licensor was not happy with that kind of job and they did not

want to give guarantee not only on performance but also on safety. On literature study, it was found that short circuit of flow can be avoided through this mechanism and at the same time, there was no alternative to carry out refinery operation. Hence, it was decided to go ahead with above two actions. Unit restarted; fortunately, unit could be run up to 94% capacity; if increased further, short circuit of hydrogen flow felt to happen. Hence, unit operation was continued at 94% capacity and that too for 8 months till fresh catalyst could be made available and refinery turnaround shutdown could be planned.

- **Lesson:** It was a troubleshooting in an extreme crisis and patch engineering and technology could not be scientifically established fully for execution; it was practically taken as a chance success. Hence, the operating crew should never drag on operation in limiting operating conditions. Licensor may agree for operation with hesitation, but it should be kept in mind that licensor always tells in advance for catalyst procurement and we ignore their recommendation failing to justify replacement as catalyst appeared to be premature as per design guarantee clause and could not contest with licensor as they find some gaps in operation procedures which was not informed to them in time.

6.5 Diesel Hydrodesulphurization Unit

Diesel hydrodesulphurization unit (DHDS) operation is one of the simplest hydrotreatment operations as compared to HCU and diesel hydrotreating unit (DHDT); its operating pressure remains at a level of about 45 bar as required for the reactions, viz., desulphurization and denitrification to happen. Some examples of troubleshooting experiences are given as follows.

6.5.1 Flash Point Off-Spec in Run Down Product HSD

- **Problem:** While all operating parameters were normal, HSD r/d flash point was found to be low; there was no scope of leakage effect in r/d exchanger; hence, stripping steam rate had to be increased though its pressure and temperature were normal.

- **Solution:** Reducing feed by 20% and increasing steam by 30% solved the issue. In the next s/d, it revealed many bottom trays got dislodged.
- **Lesson:** Always commission stripping steam after thorough condensate draining.

6.5.2 Recycle Gas Compressor About to Trip at High Flow

- **Problem:** It was observed quench flow has increased on auto temperature mode though in furnace COT and feed quality no change observed. The situation forced to reduce main recycle gas to feed.
- **Solution:** Quench c/v taken on manual mode and quench flow reduced; this helped in increasing main gas flow; problem solved. Subsequently, Instrument Maintenance rectified the quench control valve.
- **Lesson:** All flows/ temperatures should be compared with design even operated in auto control.

6.5.3 Run down product HSD hazy and emulsion water found to be present:

- **Problem:** Coalescer was not functioning.
- **Solution:** Run down HSD temperature increased from 36 to 40°C, watching tank temperature which remained within 37°C just below HSD flash point; moreover, it was floating roof tank; problem solved; coalescer design got corrected in next s/d.
- **Lesson:** Keep margin in coalescer design.

6.5.4 Frequent Problem in Feed Filter Element

- **Problem:** Delta pressure of feed filters increased rapidly and cartridge elements to be replaced as those were disposable; at the same time upstream mechanical filter delta pressure was normal and no dirt; it resulted in high filter cartridge inventory.

- **Solution:** As there was scope to increase delta pressure in the mechanical filter, its strainer size reduced to reduce load in cartridge; the problem got solved.
- **Lesson:** Study and compare design data with operation requirements and innovate.

6.6 Diesel Hydrotreating Unit

It is also a fixed-bed hydrotreatment process and operating pressure severity is in between DHDS and HCU, i.e., at about 100 bar as compared to DHDS at 45 bar and HCU at 180 bar because in DHDS only desulphurization and denitrification required which requires low pressure as above, while in HCU, in addition to desulphurization/denitrification/demetallization, closed ring saturation like thiophane, pyridine, etc., along with part aromatic saturation plus hydrocracking of saturated closed ring compounds as mentioned above are required to happen; but in DHDT, in addition to desulphurization, denitrification of open chain compounds and of the side chains of closed compounds, desulphurization and denitrification of closed ring compounds, viz., thiophane and pyridine respectively occur through hydrocracking of these closed ring compounds along with part hydrocracking of naphthenes thus helping more removal of sulphur and reduction in boiling range of the product as compared to that in DHDS. Also, sometimes mild hydrocracking of aromatic compounds occurs additionally depending upon design criteria. As conceptually, it is a similar technology like DHDS and HCU with the exceptions as mentioned above, the above explained troubleshooting experiences can be utilized in DHDT operation also; however, one specific case study is detailed below as applicable for DHDT only.

6.6.1 Sulphur Content of Product Diesel Showing Higher Than Projected in Design

- **Incident:** The unit was designed to operate at about 65 bar pressure in the reactor and product sulphur was supposed to be 8 ppm maximum; but when operated, sulphur reported higher though hydrogen make up, recycle gas flows, operating temperature, pressure, hydrogen purity, feed design quality, i.e., all were normal. The product sulphur content came down only by reduction of feed flow rate, i.e., capacity of the unit.

- **Troubleshooting:** It was taken up with the licensor and it could be understood that while fixing feed qualities in design criteria, only PONA (paraffin, olefin, naphthene, aromatic) content along with sulphur, nitrogen, boiling range, CCR and density were considered. Mass Spectroscopy analysis for the feed was not carried out. One can decide to know the nature of thiophane compound, i.e., the concentration of particular category of thiophane apparently blocked by other aromatic compounds/hetero-cyclic compounds to make it difficult for the hydrogen to reach to the sulphur of thiophane at the designed space velocity. With reduction in feed rate, as space velocity came down, this sulphur could get time to react with hydrogen to achieve product sulphur content.
- **Solution:** Existing catalyst to be replaced with the right catalyst in the next opportunity as changing operating severity may not permit to do so as limited by hardware design.
- **Lesson:** Flaws in licensing agreement key points should be the part of the tendering process to take care of the flaws / lapses appearing after placement of purchase order.

6.7 DHDS/HCU/DHDT (Common Issues)

6.7.1 Skimming of Catalyst in Top Bed of Reactor

- **Problem:** Reactor top cover flange and internal found jammed and could not be opened: tried with torque range with increased tension; not possible.
- **Solution:** Top cover two bolts had to be cut and flange thread retooling done, while internal segments could be dismantled by soaking; was possible for having spare stock of stud/bolts.
- **Lesson:** Keep stocks of spares before shutdown planning.

6.7.2 Lubricity Specification in Finished HSD in Tank Fails

- **Problem:** There was no scope of back blending with straight run gas oil to the diesel pool after implementation of Euro-IV HSD; blending with heavy naphtha tried, but the flashpoint went off. Low sulphur straight run gas oil was also not available.
- **Solution:** Lubricity additive explored; dosing @50 ppm solved the problem.
- **Lesson:** Regular procurement of lubricity additive started.

6.8 Naphtha Hydrotreatment Unit

Naphtha hydrotreatment is basically desulphurization process along with denitrification of traces of nitrogen present in the feed like in DHDS but it is simplest form of desulphurization due to very low sulphur content in feed, i.e., in the range of 50 to 250 ppm thereby requiring very low system pressure with hydrogen, say, about 20 bar. Naphtha being lighter and clean fluid, people do not face troubles like in DHDS as mentioned above. However, typical problems came across in naphtha hydrotreatment unit (NHTU) are discussed as follows.

6.8.1 Reactor Pressure Drop Increasing Unexpectedly

- **Problem:** After feed tank change over, reactor pressure drop increased suddenly, which increased hydrogen compressor discharge temperature due to consequent high compression ratio, and flow had to be reduced to reduce compressor discharge temperature. Consequently, feed also had to be reduced by 10%. Moreover, it raised concern on operation cycle length reduction.
- **Troubleshooting:** Naphtha feed sample visually checked and compared with that of sample from other tank; it was found that other tank's sample was totally transparent unlike feeding tank; immediately, feed tank changed over to other tank with transparent naphtha and it was found reactor pressure drop not only holding but also slowly falling though

it did not come back to original level. It may be mentioned that there was no mechanical filter in the unit feed pump suction because naphtha being light and clean fluid is not required, but in other refineries, licensor provided magnetic filters in the feed line to arrest dirt fines, if any, which was not proposed here. However, as previously, such pressure drop did not arise, operating people were still looking for the reason. It was found that the feed tank went for maintenance & inspection (M&I) activities and so, it was idle for about one month when it came in contact with the atmosphere; hence, atmospheric corrosion products were formed in the feed tank inside the wall which started entering inside the unit when the tank was taken in line. Hence, naphtha of this tank was routed to excess naphtha tanks for selling as naphtha, and the tank was refilled with fresh naphtha. Meanwhile, reactor pressure drop further reduced over a couple of days though marginally higher than original level and after this feed tank was taken in line, pressure drop did not rise further.

- **Lesson:** Magnetic filters in the feed line should be installed. Also, the feed tank after M&I should be thoroughly flushed before commissioning.

6.8.2 Catalyst Activity Appeared to be Decreased Prematurely

- **Problem:** During the course of operation, it was observed that reactor inlet temperature had to be increased more than expected as per duration of operation in the cycle; otherwise, sulphur content of product naphtha could not be reduced up to specification.
- **Troubleshooting:** It was found feed sulphur content had never increased; moreover, feed sulphur content in this unit was always low at about 20 to 40 ppm as compared to the allowable limit of 200 ppm. It was also found that the licensor provided DMDS (dimethyl disulphide) injection facility in the feed to keep minimum H_2S content of 50 ppm in the recycle gas to prevent leaching out of catalyst sulphur as was set by the licensor. It may be mentioned that catalyst activity remains normal when it is in sulphided form, that's why, licensor

supplied catalyst in oxide form and this oxide catalyst was sulphided in commissioning before feed cut in as per licensor procedure. Also, the operating crew has to continue DMDS injection in normal operation as mentioned above. It was also found that DMDS injection was being continued but there was no schedule of analysis of recycle gas H_2S which was done once in a blue moon; hence, nobody checked the adequacy of DMDS dosing rate. After this analysis, it was found that recycle gas H_2S was very low at 25 ppm. Hence, the dosing rate increased to maintain recycle gas H_2S of 100 ppm as allowed maximum. After leaching out of catalyst sulphur, it is very difficult to get sulphur fully re-absorbed on catalyst as catalyst is reduced to metallic stage. Hence, operation could not be continued up to desired cycle length and catalyst regeneration had to be done to get back desired catalyst activity.

- **Lesson:** Licensor recommendation on schedule of lab analysis should be strictly followed and if at any point of time it was felt to be relaxed, it is to be done taking licensor into confidence.

6.8.3 Reactor Effluent Exchanger Pressure Drop Increased Causing Back Pressure Upstream

- **Incident:** There were six heat exchangers in series with the reactor effluent in tube side and feed in shell side. It was observed that during the course of operation tube side pressure drop increased which caused back pressure to both feed pump and hydrogen gas compressor resulting in reduction of t'put.
- **Troubleshooting:** Initially, it was doubted whether any catalyst fines or internals of the reactor entered the exchanger tube side or not. But reactor delta pressure was found normal; hence, any abnormality from the reactor should not happen. It was observed that there was a DM water injection facility in the tube side; it was never required to be used. Hence, the operating manual was studied and found that DM water injection should be done during pressure drop rising in the tube side. Unit brought to shutdown and kept in a hydrogen medium. DM water injection started and washed water

drained from the downstream flash vessel. The unit restarted and found normal. Water injection in the running unit was not done from fear that in case water gets carried over with hydrocarbon due to less settling in there, there might be water shot in the next column, i.e., stripper.

- **Lesson:** Sometimes in operation, feed sulphur content went high which caused ammonium bisulphide deposit in tubes. Hence, feed quality should be religiously monitored to avoid crisis.

  Note: this case study was also discussed in the exchanger section; hence it's a repeat.

6.9 Catalytic Reformer Unit (CRU)/Continuous Catalytic Reformer (CCR) Unit

It is a reformer unit for catalytic reforming of light naphtha, after the naphtha is hydrotreated as mentioned in section 6.8, to produce motor spirit (MS) of desired research octane number (RON) at about 98 in case of CRU and about 105 in case of CCR. In CRU, the catalyst used is of semi regenerative type, i.e., operation can be continued up to desired cycle length, thereafter it should be regenerated in situ. To avoid unit shutdown for in situ regeneration of catalyst, for two decades, continuous catalytic regeneration (CCR) system has been introduced in the industry by the licensors which is costlier in capital expenditure (CAPEX) as compared to CRU. However, operation of CRU is simpler, i.e., with much less instrumentation and security interlock system and without additional equipment as required for CCR process. Till today, CRU is in use in many refineries, troubleshooting in CRU also has been discussed along with those in CCR as follows.

6.9.1 Unexpected Decrease of RON, Probable Reasons

- Single or multiple possibilities:
  - Leak in the feed/effluent exchangers.
  - Presence of nitrogen in the feed.
  - Presence of sulphur in the feed.
  - Presence of metals in the feed.

- Low chloride content of the catalyst.
- Partial by-pass of the catalyst.

6.9.2 Loss of Reformate Yield, Probable Reasons

- Single or multiple possibilities:
  - Presence of sulphur in the feed.
  - Presence of metals in the feed.
  - Too high chloride content of the catalyst.

  Recommendation: Similar actions to be taken as explained in the recommendation against section 6.9.1. Simultaneously, it should be looked into whether quality of naphtha feed in NHTU is inferior or not; if yes, then same to be improved by improving crude mix or eliminating contamination in crude mix due to slop or some contamination in NHTU due to mix up in line up or malfunction in NHTU process.

6.9.3 Unexpected Delta Temperature Decrease in First Reactor, Probable Reasons

- Single or multiple possibilities:
  - Feedstock quality: naphthene dehydrogenation is highly endothermic. Lower naphthene content in feed is reflected immediately on the ΔT in the first reactor.
  - H_2/HC ratio increase (increased recycle). An increased recycle acts as a larger heat sink for the heat of the reaction.

  Recommendation: New set of operating conditions are to be devised with the help of process licensor.

6.9.4 Other Reasons for Delta Temperature Reduction Across Reactor Beds

- Single or multiple possibilities:
  - Presence of sulphur in the feed.
  - Presence of nitrogen in the feed.
  - Too high chloride content (which favours hydrocracking—exothermic reaction).

- **Recommendations:** A loss of ΔT in the first reactor generally results from sulphur upset in the unit feed. Poor operation of the pretreatment reactor or of the stripper can then be suspected. Severity must be reduced, according to the H_2S content in the recycle (for H_2S > 5 ppm volume, reduce temperature to 480°C).

  In case of low ΔT in reactors 2 and 3, hydrocracking can be suspected. If such is the case, H_2 purity drops and LPG production increases. But the most worrying aspect of a decreased ΔT, when it cannot be related to one of the above grounds, is the result of a permanent poisoning with metals.

6.9.5 High Hydrocracking Rate

- Probable reasons:
  - A fresh catalyst (too active) or an excessive acid function of the catalyst (i.e., high chloride).
  - A too high severity (either too high temperature at the reactor inlet or too low space velocity).
  - Poor distribution (channelling) in the reactor which results in local high severity conditions.
- Recommendations:
  - Not to operate at capacity below the design turn-down ratio (generally 60%).
  - Not to introduce feed on a fresh or regenerated catalyst before sulphiding.
  - Not to introduce feed for the first time on a fresh or regenerated catalyst at temperature higher than 400°C. For a subsequent restart, then the temperature limit should be increased to 430°C.
  - When raising throughput, to raise flow first and increase temperature afterwards.
  - When lowering throughput, to lower temperature first and throughput afterwards.
  - To raise temperature always do cautiously: proceed by increment of 2°C/h.
  - To monitor the differential pressure in the reactors. An increased differential pressure is the sign of poor

distribution of feed in the reactor if cycle duration is not high.

Note: If unexpected higher rate of hydrocrackings suspected, then do following alternatively:

- Lower the reactor inlet temperature down to 400°C.
- Increase the feed rate to full capacity (to increase space velocity).
- Keep (or set) the recycled gas flow to the maximum.
- Consider catalyst dumping after next regeneration.

6.9.6 Delay in Shutdown Period in CRU

6.9.6.1 Case I

- **Incident:** A semi regenerative type reformer was shut down for catalyst regeneration. The unit was shut down and evacuation/pressure purging was done to make the high pressure system H_2/HC free. The refinery did not have N_2 plants. N_2 was being supplied in cylinders. There was a N_2 bullet having a capacity of 150 M3. Supply of N_2 used to be 6 cylinders on alternate days or even higher. Once the system was made free of Hydrogen plus Hydrocarbon at <0.5%, N_2 was taken and a compressor was started. Subsequently furnaces were lighted for first phase regeneration.

  Before air introduction, samples were again checked and the explosive limit was found higher.

  Again units brought down and purging done to bring the limit to specified level. After the first phase temperature was stabilized the samples again showed higher value than last time. Again the system was brought down but before air introduction, the H_2/HC value was found to be still higher and even liquid drops were coming from the reactor effluent cooler outlet. The value reached 11% by that time. There was a severe shortage of N_2, due to prolonged s/d period. Also, consequently due to delay in hydrofinishing unit start up for want of H_2 from CRU, the lube block was under threat of S/D.

- **Troubleshooting:** It was seen from operating manual that after removing hydrocarbon in cold condition, when regeneration to be started, gas sample need not be checked against hydrocarbon content before air injection; but this was being done by mistake and naturally in hot condition the entrapped oil inside catalyst pores was coming out and thus getting reflected in gas analysis and misleading the whole operating crew. Hence, finally this time no gas analysis was done at hot condition before air injection and no problem observed in regeneration process and regeneration completed successfully without any fire, explosion or any problem.

- **Lesson:** SOP for regeneration was not followed; rather people became extra cautious going beyond operating instruction.

6.9.6.2 Case II

- **Incident:** In another refinery, the reformer created a record by running for more than 8 years without shutdown. While running for such a long period, no problem was faced. Except for checking RON, no other checks were there for any stream. The unit then planned for a regeneration shutdown. A committee reviewed the procedure and planned for 15 days regeneration and another 7 days for maintenance. After shutdown, the plant remained under shutdown for months. During shutdown, the recycle gas compressor could not be run due to lots of maintenance issues.
- **Troubleshooting:** In spite of objections to plan such short shutdown bypassing many checks, the plan was finalized by the planning group. The avoidable delay was due to ego, not listening feedback from experienced persons, and lack of knowledge about reformer operation.
- **Lesson:** For future planning it was decided that compressor maintenance also should be included in the shutdown schedule if the same run for a long period without maintenance.

6.9.6.3 Case III

- **Problem:** In another case during startup and sulphiding stage, first stage sulphiding was done in night shift and second stage was to start in M/S (morning shift). As three step DMDS dosing is time bound, for each step delay or slow/fast dosing is not permissible. It was found that the injection line to the third reactor was choked and no dosing was possible. No alternate guideline was there. This was delaying the process.

- **Troubleshooting:** Dosing was done through a furnace normal line by connecting flexible hoses with proper connection and the problem got solved.
- **Lesson:** Sometimes operation has to survive with an ad hoc solution, but a HAZOP study of ad hoc process also should be done even if it delays the job.

6.9.7 Chokage in Regeneration Circuit in CCR Operation

- **Problem:** In one of continuous catalytic reformer unit (CCRUs), there used to be a frequent choking problem in the regeneration circuit due to catalyst fines. The regeneration used to be bypassed and after cleaning it was being taken online; the catalyst loss also was on the higher side. The catalyst sample was showing chloride within range. During startup, the temperature normalization was being done based on the catalyst chloride level. But this modified procedure in crisis affected performance of the unit.
- **Troubleshooting:** The problem was created due to not following recycling gas moisture level maintaining guidelines.
- **Lesson:** It was a simple example of violation of licensor's procedures/instructions.

6.9.8 Poor Performance Splitter Reboiler in CRU

- **Problem:** In one CRU, the naphtha splitter reboiler was failing very frequently. During regeneration shutdown, the reboiler was taken for maintenance. Inspection had recommended a tube bundle change. Gummy type brownish deposited used to be found out every time when t/b used to be opened. In crude distillation unit (CDU) there was a facility of naphtha caustic wash prior to sending to reformer.
- Probable reasons:
  - In CDU naphtha wash was not having water wash.
  - High t'put made partial bypass

  - Caustic carry over to tank and to splitter
  - The gummy material tested for pH, found highly alkaline (caustic).
- **Lesson:** There was contamination in desulphurized naphtha tank which was used during start-up of CRU after shutdown. SOP should be developed to control the quality of naphtha in that tank.

6.9.9 Instrumentation Problem in CCR

- **Problem:** In one CCRU, it was found in the evening shift that suddenly the r/d flow of reformate became zero. The plant was absolutely steady and normal just before the problem.
- **Troubleshooting:** Run down control valve stem got detached and closed.

6.9.10 Catalyst Chlorine Leaching Out in CRU

- **Problem:** In one CRU, RON was lower than expected in the middle of the cycle. H_2/HC mole ratio was normal at 5.5; hydrogen purity was marginally lower at 87% than expected 88%. Recycle gas moisture level was at 25 ppm which was just required for normal operation; recycle gas chloride content was lower at 0.6 ppm than expected of 0.8 ppm. Reformate RON was lower at 96 to 97 than expected.
- **Troubleshooting:** It was observed in the refinery that many of the on-line analysers were giving wrong reading due to absence of regular monitoring of the instruments and inadequate heat tracing of the impulse lines. For recycling gas moisture content analysis, laboratory methods like bringing the instrument in the field and analysing in there was also not true representative; hence, field lab analysis also not giving reliable value.

  In this situation, previous lab reports of catalyst carbon and chloride content were seen; it was found that both chloride and carbon contents were normal. As carbon content was not high, doubt on recycle gas moisture content further strengthened and it was decided to increase water doing with monitoring on recycle gas chloride content so that when recycle gas chloride content went down to 0.6 ppm, chloride dosing started and when chloride level went up to 0.8 ppm chloride dosing stopped. With this action, it was observed that every day RON was increasing and finally reached the expected level of 98 whereas recycle gas hydrogen purity did not deteriorate from its existing level of 87%. This was done as there was a scope to increase the carbon content level in the catalyst in both first and second stage; however, carbon content in the third stage catalyst had only marginal scope. Hence, catalyst carbon content analysis was done as per schedule religiously.

- **Lesson:** In addition to fixed chlorine in the catalyst, recycle gas chloride level also should be maintained which was not followed strictly resulting in marginal leaching out of catalyst chlorine. This situation could be neutralized by the action as mentioned above.

  Exception: In this process, if there is by chance excess chlorine slippage to recycle gas, the same gets removed through chlorine guard before routing to PSA unit of hydrogen generation unit (HGU) for subsequent routing to all hydrotreating units.

  But in one instance in one CRU the catalyst activity was severely affected due to one fire in that unit. To sustain operation till fresh catalyst procurement, similar chlorine dosing was started but at higher rate to get MS RON. Also, there was no PSA unit for hydrogen purity improvement in that refinery as there was no HCU in the refinery. The hydrogen purity ex CRU was adequate for NHTU and KHDS (Kero Hydro-Desulfurization) unit. In NHTU, there was no makeup compressor but in KHDS unit there was make up compressor which after two months operation had to be taken to maintenance as discharge temperature went high; it was found that there was huge ammonium chloride deposit in the loader valve which was due to this excess chlorine in recycle gas as mentioned above.

  However, fete accomplice, such a way of operation was continued with alternate operation of standby compressor till six months when fresh catalyst was made available.

6.9.11 Capacity Limitation in CRU due to Gap in RG Compressor Design

- **Problem:** In CRU, t'put was always maintained at 95% max as there was no extra feed. When available and t'put increased, recycle gas compressor speed was also increased to maintain H_2/HC mole ratio unchanged; but the spill back control valve of the compressor was getting open thus restricting t'put increase.
- **Troubleshooting:** When analysed, it was found that at that level it should not have happened. From study of the data sheet, it was discovered that design molecular weight of recycled gas was kept higher by communication gap in the design stage. Compressor manufacturer was consulted; they agreed to execute debottlenecking in the next shutdown, and for the time being it was decided to increase t'put without increasing recycle gas compressor (RGC) speed thus sacrificing H_2/HC mole ratio up to 5.0 from 5.5 as per catalyst carbon content analysis which indicated unit could be operated till end of cycle.
- **Lesson:** Such a problem surfaces if guarantee test run is not done properly after unit commissioning.

  Note: The case has also been discussed in compressor section detail.

6.9.12 Feed Availability Problem in CRU

- **Problem:** In CRU, MS market demand increased but adequate feedstock was not available even after operating CDU at design t'put. Further increase of CDU t'put became critical thus restricting further MS production increase.
- **Troubleshooting:** As per final boiling point of motor spirit (MS FBP) specification limit, there was enough scope to make CRU feedstock heavier, but in that case CRU catalyst carbon content would increase faster resulting in reduction in cycle length. On lab analysis, it was found that the carbon content in the catalyst of the first and second reactor was low, but in the third stage catalyst there was not much scope. Hence, it

was decided to increase naphtha yield in CDU by pulling some heavy naphtha into it and operating first and second stage severity in CRU by 1°C. This helped in increasing CRU t'put by 2 m^3/h and CRU operation could be continued till the next planned regeneration shutdown.

- **Lesson:** This type of decision to operate units beyond standard guidelines should be followed involving higher ups in the department/organization.

6.9.13 Feed Quality Limitation in CCRU

- **Problem:** In CCRU, reformate RON was lower than expected; all operating parameters were normal. On lab analysis, it was found actual feed composition was different to some extent as compared to design.
- **Troubleshooting:** All scopes to increase RON were explored; it was found that by reducing RGC discharge pressure, i.e., reducing system pressure slightly RON could be increased by 1 unit and reach level of expectation as explained below:

  In CRU/CCRU, Hydrogen is a product and we know that reaction rate in terms of product concentration is given by:

  $$R = K/(x_H \cdot P)^n, \qquad (6.2)$$

  where K is gas constant; P is total pressure of the reactors; x_H is the mole fraction of hydrogen in the product, and n is the order of kinetics of the reaction.

  From above, it is seen that by increasing the value of P, reaction rate increases. To utilize this scope, a problem in downstream section hydrogen rich compressor allowable suction pressure limit was checked and found there was scope to do it.

  Hence, by utilizing the scope of increasing total pressure, reformate RON could be increased even with minor limitations in feed quality.

  Note: Mind it this concept can be applied in CCRU but not in CRU because reducing partial pressure of hydrogen as above, cracking reaction to form coke also occurs which to some extent can be allowed in CCRU being of continuous regeneration mode unlike CRU where regeneration is done in situ after a completion of operation cycle thus cycle length of operation would come down in case of CRU.

Lesson: Technical service/planning department should be highlighted to take up corrective action in crude procurement.

6.10 C5 Naphtha Isomerization

C5 naphtha is basically light naphtha and cannot be used as feedstock in CRU for MS production. Earlier, it was blended in an excess naphtha tank and was sold as a product, naphtha. But for more than a decade, the refinery installed isomerization unit (ISOM) to boost up its RON from level of 70 to level of 85 and above. It is a simple fixed-bed catalytic conversion plant with objective to convert normal paraffin to iso-paraffin having higher RON in hydrogen pressure though in the reaction hydrogen is not consumed. Negligible hydrogen used as make-up is released from the system along with off gas.

The refinery ISOM unit was licensed by AXENS, France. In that unit, for the isomerization reaction to happen in the reactor, platinum-doped catalyst was used which got poisoned in presence of Benzene. Hence, another reactor upstream is installed to saturate this Benzene; this reactor is called Benz-sat reactor. The catalyst used in Benz-sat reactor also got poisoned in presence of sulphur. Hence, desulphurized naphtha is used in this unit as feed.

Hence, this unit is connected in series with upstream NHTU for removing sulphur from it and also where desulphurized naphtha is split into C5-light naphtha for use in ISOM and heavy naphtha to use as feed in CRU/CCRU.

6.10.1 Non-performance of Benz-Sat Reactor after Start Up

- **Problem in the unit:** The problems were found usually common as discussed in other hydrotreatment units as discussed above, but there was unique problem encountered in this unit time to time as follows:

  Whenever total refinery shutdown or NHTU shutdown occurred, naturally ISOM unit also came under shutdown. Next when NHTU started up again, subsequently after stabilization, ISOM unit also was started, but in ISOM unit it was observed that in Benz-sat reactor desired exothermicity not achieved indicating that catalyst got poisoned by sulphur as no other reason possible for fixed feed source as mentioned by licensor. But every time it was puzzling for the operating crew to agree that sulphur slippage occurred from NHTU as they did feed cut in after testing product C5-light naphtha for sulphur content.

- **Troubleshooting:** In troubleshooting, different time different reason found as follows:
  a. Before start-up unit hold up was not totally flushed out.
  b. Sulphur guard in NHTU was withdrawn and bypassed after getting NHTU product with respect to sulphur content but before stabilizing ISOM.
  c. Slippage of off-spec light naphtha to ISOM unit before its start-up through battery limit valve passing.
  d. While flushing out ISOM units hold up after emergency shutdown, some short circuit of off-spec naphtha to Benz-sat reactor occurred through valve passing.
  e. Passing of sulphur guard bypass valve to ISOM unit before NHTU stabilization.
- **Lesson:** As seen above, as a whole it was mostly a valve passing problem. Hence, in ISOM, one should check reliability of valves not only before start-up but also always as it can happen during emergency shutdown by valve passing.

  Note: Unlike this AXENS' licensed ISOM unit, the ISOM units licensed by UOP or GTC do not consider platinum-doped catalysts rather zeolite based and titanium sulphate/phosphate/silicate based catalysts, respectively. However, in zeolite based catalyst reaction temperature has to be kept on the higher side with higher hydrogen recycle ratio unlike AXENS process thus with higher energy consumption. The GTC catalyst as mentioned above requires in between severity. But AXENS process needs continuous chloride dosing to maintain catalyst acidity whereas other two types of catalyst have improved acidity inherent in the catalyst thus no chloride dosing is required thus troubles with respect to corrosion also does not arise in these two alternative technologies.

6.11 Propane Deasphalting (PDA) Unit

This unit is used for extraction of de-asphalted oil (DAO) and separation of asphaltene from vacuum residue (VR) with the objective to use DAO either in FCCU for fuel distillate production or in the furfural extraction unit (FEU)/N-methyl pyrolidyne (NMP) extraction unit (FEU/NMPU) followed by solvent de-waxing unit (SDU) and hydro-finishing unit (HFU) for lube oil base stock (LOBS) production. The solvent used to extract DAO from VR is liquid propane; the liquid propane also acts as anti-solvent to asphalt.

For use in FCCU, the specification for DAO is generally relaxed to some extent, viz., it can be taken at high viscosity as well as with high CCR of about 4.0%; the % addition of this DAO to FCCU feed blend depends on its metal content and CCR. As such, there was no troubleshooting case came across in FCCU for use of DAO in feed blend. However, DAO quality is very important if it is to be used for LOBS production. Hence, to maintain DAO quality particularly, troubleshooting in Propane de-asphalting unit (PDAU) is discussed as follows.

6.11.1 DAO Yield Low

- **Problem:** For using DAO in lube processing units, its most important quality criterion is CCR which should be 2.5% max. Considering 2.5% as ideal value of CCR, DAO potential in VR should be calculated as:

  $$((\text{CCR of asphalt-CCR of VR})/(\text{CCR of asphalt-CCR of DAO})) \times 100\ \%\ \text{wt} \quad (6.3)$$

  Hence, with asphalt CCR of about 25 and VR CCR of about 18 (avg. Value of VR CCR from different crude sources), DAO potential yield comes about: 31% wt but this is based on 100% phase separation between DAO and asphalt. In actual, there is some cross-mixing between these two phases; hence, achievable separation should be 95%, i.e., DAO yield should be: 31 × 0.95 = 29.5%.

  But in reality, yield came as low as 25%, which ultimately caused lower production of Bright stock Lube Base Oil whose feedstock is DAO.

- Probable causes identified at different times are:
  a. Lower solvent (propane) ratio at the extractor(s) in PDA unit though as per design solvent ratio was maintained. In such a case, VR CCR was analysed and the above calculation was done and DAO potential was to be higher than actual achieved. As to get extraction efficiency, design solvent load/flow rate to the extractor was optimum; instead of increasing solvent ratio, stand by extractor should be taken in line by reducing t'put per extractor thus getting higher solvent ratio per extractor and thus increasing overall solvent load to the unit as far as DAO-Solvent recovery circuit capacity permits.
  b. Extractor top or bottom temperature or both were high. In that case also, DAO yield would be low as propane solubility to oil decreases with increase in temperature. But in that case DAO CCR should be better, i.e., lower than 2.5%. If DAO CCR is low, then extractor temperature should be reduced to get back the DAO lost through asphalt; but if DAO CCR was ideal, then the extraction efficiency is low and, in that case, additional extractor is to be taken in line when high solvent ratio can be afforded per extractor.
  c. Sometimes feed, VR temperature at an offsite tank is very high in which case even after closing steam to top steam heater in the extractor, its top temperature remained high; in that case, feed tank steam coil heating should be stopped only during its processing if upstream unit could not reduce VR run down temperature.

6.11.2 DAO CCR High

- **Problem:** It has been found many a time that DAO CCR became higher than 2.5% though actual yield was still lower than potential yield. It is known that DAO with higher CCR when processed through aromatic extraction unit, its CCR though gets reduced but still remains at higher level which is detrimental to the next processing unit, SDU where bigger size asphaltene compounds clog the filter cloth thereby reducing production rate. Hence, CCR should be reduced in the PDA unit as discussed below.

- Probable causes identified at different times:
  a. Solvent (Propane) ratio is too low due to some equipment problems. If solvent ratio is marginally low, then DAO yield would be low but its CCR also should be low; but solvent ratio is too low, then there would not be proper mixing between Propane and VR thereby effecting short circuit of VR into DAO thus DAO CCR would be high. So, there should be first optimum solvent ratio, then extractor top temperature should be increased to increase temperature gradient between top and bottom to the level of 22 to 25, an ideal gradient for optimum extraction. There is a steam heater at the extractor top where temperature is increased to get the above.
  b. Sometimes top recycle, i.e., circulating a part of top section DAO back to down section of extractor helps to reduce DAO CCR. But it was found that the recirculation line remained choked and generally people do not put impertinence to that as it gives effect as a mode of fine tuning of DAO. But if the line gets choked it affects badly on DAO CCR; hence, steam tracing effectiveness in this line should be strengthened.
  c. Sometimes in feed (VR), CCR is higher than 20%; in that case, DAO potential yield should be low; hence, to get DAO CCR within range its actual yield cannot be increased beyond its theoretical yield.
  d. Propane purity is lower than design; thus DAO CCR would be higher as butane present as contaminant in propane would dissolve some asphaltene into it thereby increasing CCR of DAO; this time DAO yield also becomes high; thus some people decrease the solvent ratio to reduce the yield but CCR of DAO is not improved as solvent ratio became too low affecting homogeneous mixing of solvent and VR as mentioned above thereby affecting CCR of DAO. So, propane purity should be maintained. If purity cannot be maintained, then DAO CCR should be further increased up to 4% and it should be routed to FCCU instead of routing it to lube processing units.

6.12 Furfural Extraction Unit (FEU)/NMP Extraction Unit (NMPU)

These two units are used to remove aromatics from vacuum distillates for subsequent processing in solvent dewaxing unit (SDU) for producing lube oil base stock (LOBS). As in FEU, furfural degradation occurs with consequent increased make up of furfural in the system; most of the new extraction unit is designed to use NMP as solvent as NMP degradation is less as compared to that of furfural while selectivity is comparable. The major problems encountered in the unit are discussed as follows.

6.12.1 Both Raffinate (Aromatic Extracted Distillate) Yield and Quality Not Achieved in FEU

- **Problem:** It happened after M&I shutdown; the unit was started with topping up of furfural which was sent back to the tank during shutdown and also freshly made up. After feed cut in, it was observed that both raffinate yield was low and quality was poor, i.e., density of raffinate was high and close to feed density; similarly extract density also was low which should have been higher than feed density; as such phase separation was not taking place. All conventional methods of improving raffinate and extract quality were tried, viz., extractor top temperature increase, furfural ratio in feed was increased by taking conventional minor make up increase of furfural in the system, but no improvement observed. One of the parameters on adequacy of furfural inventory in the system was extractor interphase level; but as most of the time interphase level indication did not appear while unit performance was satisfactory, the same was ignored.

  This incidence caused upset in raffinate production thereby affecting LOBS production.

- **Troubleshooting:** As nonappearance of interphase indication was ignored, extractor pressure indications at top and bottom were noted; on calculation it was found that differential pressure across top and bottom of the extractor was very low as compared to design which was a reflection of the fact that furfural inventory inside extractor was miserably low as explained below:

$$\Delta P = P_B - P_T \tag{6.4a}$$

$$= (h_B \times \rho_B - h_A \times \rho_A)g, \tag{6.4b}$$

where ΔP is differential pressure between bottom and top of extractor; P_B and P_T are bottom pressure of the extractor and top interphase bottom tapping pressure of the extractor respectively; h_B is the total height of the extractor and h_A is the height of the pressure tapping located at the bottom of top interphase; ρ_B and ρ_A are the densities of the extract mix sample collected from the bottom of the extractor and raffinate mix sample collected from top of the extractor respectively.

Due to ρ_B being lower and ρ_A was higher than normal value, ΔP was lower. This can happen due to poor furfural inventory when aromatic separation would not occur thereby some aromatic remaining in the raffinate phase while some raffinate remains in bottom extract phase. Hence, it was doubted that abnormal loss of furfural occurred during unit shutdown and recovered furfural in the tank during shutdown was contaminated with oil instead of pure furfural. Also, it may be mentioned that not only lower density is a reflection of good raffinate quality but also right delta density between raffinate and extract as discussed above; when this delta density is very low, there is no point of fine tuning of operating conditions to get the raffinate quality and yield, and instead other causes should be explored like explained as above.

So, it was decided to procure furfural equivalent to 70% of extractor inventory; it was also decided not to consume furfural till the whole quantity arrives; after taking whole quantity of furfural within a day, immediately raffinate and extract qualities started improving and the problem got gradually solved.

- **Lesson:** Priority attention should be given to prevent loss of solvent to avoid subsequent crisis.

6.12.2 Furfural Loss Increase in FEU

- **Problem:** Make up of furfural increasing in FEU causing increase in cost of production.
- **Troubleshooting:**
  a. In furfural quality specification, 99% purity was criterion but there was no mention against any impurity which was not acceptable. Also, it was unknown to refinery people about the impurities which might cause degradation of furfural during the course of operation. However it was known that furfural in the presence of oxygen slowly degrades. Hence, inert gas blanketing in a furfural storage tank was provided and purity criterion was increased from 99% to 99.8% which one supplier agreed to meet. After taking these actions, the problem gradually solved.
  b. Even with the above actions, sometimes, furfural consumption did not reduce that much. It was known from literature that furfural coke up with high temperature in furnaces above 230°C and this was observed also during furnace decoking in shutdown. Though furnace outlet temperature was maintained at 228°C as per standard operating instruction, furnace tube metal skin temperature naturally went above 230°C. Hence, temperature was reduced to 225°C when tube metal skin temperature was about 228°C; to maintain furfural recovery unchanged, system pressure could be marginally decreased. This reduced furfural degradation and solved the problem.
- **Lesson:** Sometimes, prevention clause is not adequately defined in the design; in that case, experience, study and innovation help to solve the problem.

6.12.3 Raffinate Yield Lower in NMPU

- **Problem:** In this case, raffinate quality was achieved but yield was low as per design and previous performance. It was found that crude-mix was not changed and also in the feed tank there was no possibility of presence of distillate from other crude sources. This caused reduction of LOBS production.

- **Troubleshooting:** As a part of quality achievement though raffinate density was low but density difference between raffinate and extract was lower than expected which was an indication of raffinate slippage to extract. Naturally, in this case, people could not reduce temperature to increase raffinate yield as this caused lifting of extract to raffinate thereby deteriorating raffinate quality.

  But how this unique situation arose; it was observed that water injection to solvent was more than stipulated as the recovery system was overdesigned to take care of pressure load by water vaporization. Water is an anti-solvent to facilitate extraction thus could be used in absence of adequate NMP flow, but generally designer does not give much scope for water addition because then cost of the recovery circuit would increase; but however, here, people could increase water injection due to scope in design in recovery circuit as mentioned above. In this way, the operating crew could achieve raffinate quality but right phase separation between raffinate and extract could not be ensured thereby losing raffinate yield. It was also observed that extractor interphase level remained at zero most of the time. Hence, it was decided to procure more NMP and add to extractor till extractor interphase level could be maintained at 30% on a sustained basis; there was no loss source due to abnormality in operating conditions. After achieving the interphase level as above, raffinate yield improved without deterioration of its quality and problem got solved.

- **Lesson:** People ignore the main problem if there is some way out though not up to full satisfaction. Maintaining all operating conditions does not create such a situation.

6.13 Solvent Dewaxing Unit (SDU) for LOBS Production

The raffinate from FEU or NMPU as mentioned above enters this unit as feedstock for chilling and filtering out the waxes from it to improve its congealing property, viz., pour point. To facilitate fluidity for the chilling and filtering process and to ensure good phase separation between wax and oil, solvent like methyl iso butyl ketone (MIBK) or mixture of methyl ethyl ketone (MEK) and toluene is used. As solvent is used, naturally there would be two solvent recovery circuits, one for dewaxed oil (DWO)-solvent phase and other for wax-solvent phase. To facilitate chilling there should be a refrigeration circuit and for efficient filtering there should be a vacuum compressor circuit with inert gas blanketing in the filter so that instead of air inert gas can enter into filtrate.

With so many complications in the process as mentioned above, there would be inherent problems encountered during operation which are tackled with good troubleshooting practices. Some major problems and troubleshooting in this technology are shown as follows.

6.13.1 Low DWO Yield, Low T'put and High DWO Pour Point

- **Problem and troubleshooting:** For LOBS process generally 50:50 mix of MEK and toluene is used as solvent because while MEK acts as anti-solvent, i.e., it throws away wax during crystallization, toluene dissolves the oil; thus in two phases, viz., one phase with oil-solvent rich in toluene and other with wax-solvent rich in MEK are formed during crystallization process. Good phase separation is governed by following formula:

$$CR_{\text{wax}} = K \times C_{\text{wax}} \times RT/(t \times \mu_{\text{f}}), \tag{6.5a}$$

  where CR_{wax} is the rate of crystallization of wax; C_{wax} is concentration of wax in the feed; RT is the residence time for crystallization; μ_{f} is the viscosity of feed; t is chilling/crystallization temperature and K is the constant.

  If more solvent is used, concentration of wax decreases thereby affecting crystallization rate as in the above equation resulting in less DWO yield; on the contrary if less solvent is used then some oil would remain in wax phase due to poor phase separation which also causes loss of DWO yield. Similarly, if more residence time is used, i.e., chilling rate is high, more crystallization would take place as evident from the above equation but in that process t'put would come down for a fixed designed series of chillers because more residence time means less t'put. For higher viscosity of feed, solvent ratio to be maintained high to facilitate better filtration but again concentration of wax would decrease resulting in lower crystallization rate as per above equation. Also, with higher solvent ratio, wax crystal sizes would be smaller thereby reducing porosity of wax cake thereby affecting filtration rate because filter drum speed cannot be increased beyond optimum for which it is built in the design of filters.

We can understand from above that for each quality of feedstock (like its wax content and its nature, i.e., whether micro-crystal or macro-crystal and high CCR of feed, i.e., with more asphaltene molecules clogging the filter cloth, viscosity of feedstock), and with fixed design of hardware like chillers' sizes, refrigeration compressor circuit, inert gas vacuum compressor circuit, filters circuit and solvent recovery circuit, one has to establish optimum solvent ratio, optimum chillers outlet temperatures, filters' rotation speed to achieve designed unit t'put and desired DWO yield. Sometimes, a higher solvent ratio brings up some wax into the oil phase due to wax sizes becoming micro particles thereby affecting its pour point (increase in pour point of DWO). Some case studies are discussed as follows:

6.13.1.1 Lower t'put

DWO yield was satisfactory but t'put was low as filters were limiting: It was found feed viscosity was normal but CCR was a bit higher. In this case, in the upstream FEU or NMPU, extraction was further improved to get lower CCR of Raffinate; if no scope found in FEU/NMPU, then in VDU, respective distillate CCR improved by avoiding overlap of heavier distillate into it. Sometimes, the chilling rate profile across the chillers in series had to be optimized to achieve desired crystal sizes having good porosity for better filtration to get desired t'put. Sometimes it was established that with excessive solvent ratio, wax crystal sizes decreased thereby reducing wax cake porosity it affected filtration rate and in turn unit t'put as explained below:

$$F_R = K \times R_F \times \alpha/(\mu \times Th_C), \quad (6.5b)$$

where F_R is filtration rate; K is constant; R_F is the rotation speed of the filter; μ is the viscosity of feed-solvent mix entering the filter; α is porosity of filter cake; Th_C is cake thickness.

Filtration temperature, t was already decided to get good crystallization as shown in Eq. 6.13a above and cold wash spray temperature to filter was maintained same to avoid change in the value of t. Though wax cake thickness was also decided by the crystallization rate as in Eq. 6.13a it can subsequently be decreased by excessive cold wash flow or increase in cold wash temperature in the filter. Cake thickness nowadays is measured by installing an online thickness probe in the filter. R_F was maximized up to allowable speed. By increasing solvent dilution ratio, μ was also kept decreased which should help in getting higher filtration as seen from above Eq. 6.13b, also, higher dilution ratio decreases the cake thickness, Th_C which also helped in getting higher filtration rate. But, with excessive solvent dilution rate wax cake porosity decreases. Hence, solvent dilution ratio should be kept optimum to balance good crystallization as well as good filtration rate, and at the same time, one should not forget that if cold wash temperature in the filter deteriorate, i.e., increases then wax crystals would be melted thereby reducing porosity of the wax cake. DWO pour point also increases due to increase in cold wash temperature or due to excessive solvent dilution ratio.

In this case, the problem occurred due to excessive solvent dilution ratio which after optimization, the problem got solved.

6.13.1.2 DWO yield was low but unit t'put was normal

Solvent ratios were low or filter speed low or number of filters in line less; first trial to be done by taking additional filter (s) in line if available otherwise by increasing filter rotation speed and sometimes by increasing cold solvent wash spray in the filters if it is less. If these trials do not work, then the solvent ratio should be increased. Regarding solvent ratio increase, it should be noted that solvent is added at three locations, first addition called primary dilution is done at feed pump discharge, second addition called secondary dilution is done after bulk crystallization already taken place, i.e., it should be added at outlet of first chillers' train, so that in further crystal growth in second and third chillers trapped oil in primary crystals are taken out to improve DWO yield. However, sometimes third solvent addition called tertiary dilution is done at the outlet of the last chillers' train (third chillers' train) to reduce viscosity of wax-oil slurry to enhance filtration rate thereby helping in increasing t'put. Chilling rate is also important. Generally, chilling rates of 2°C/min in the first 2 chillers in the series and 4°C/min in the third (last chillers' series) to maintain desired crystallization.

6.13.1.3 Both DWO yield and t'put were low

Either adequate numbers of chillers and/or filters were not available or desired chilling rate as mentioned above were not maintained and/or either inefficient functioning of refrigeration compressor and/or its control circuit in chillers. Sometimes with very low solvent ratio, this can happen but, in that case, DWO pour point would be too low, i.e., quality giveaway occurred.

6.13.1.4 DWO pour point high:

This happened when final chillers' outlet temperatures went high or sometimes with very high solvent ratio as mentioned in this section above. Sometimes, when cold wash spray temperature in filters went high, this happened, but in that case, ultimately filter cakes would get damaged if temperature went too high thereby reducing filtration capacity ultimately resulting in unit t'put reduction.

Note: Snapping of wires of filter cloth winding of course reduces filters availability and unit t'put, but also sometimes this caused tearing out of filter cloth thus affecting DWO pour point through short circuit of wax into DWO; sometimes poor filter cloth quality and/or abnormal hot wash of filters for its filtration efficiency regeneration caused tearing out of cloth and thereby affecting DWO pour point.

Common lesson: SDU operation is practically an art, and experience is earned by trial and error based on the above backgrounds.

6.14 Catalytic Iso-Dewaxing of Vacuum Distillate or Raffinate

This process came into effect two decades ago. Here, all waxes are converted into oil by isomerization and aromatics get saturated by hydrotreating thereby improving viscosity index (VI) property of LOBS. Hence, instead using raffinate as feed, here, straightway, vacuum distillate can be used. However, where extraction units were already in place, there catalytic iso-de-waxing (CIDW) unit of lower severity has been installed as aromatic content in raffinate is less than in vacuum distillates.

In this process, there are series of reactors where in first reactor hydrotreating/hydrocracking takes place to improve VI of LOBS, in second reactor isomerization takes place where waxy paraffins get isomerized to become non-waxy thus facilitate to achieve very low pour point of DWO. However, in isomerization reactor some napthenes gets dehydrogenated back to aromatics; hence isomerization reactor outlet goes to third reactor for mild hydrogenation to get back lost VI in second reactor.

In CIDW process, paraffin content of the finished Lube base oil increases to/above 97%; hence, DWO obtained through this process is called group-II LOBS even with –6°C pour point. When DWO pour point with 97% min paraffins reaches –9°C pour point in this process, then Lube Base oil it is called group-III LOBS. In CIDW process it is easy to produce group-III LOBS in addition to group-II LOBS. However, sulphur limit is 500 ppm, Max for both the categories which is easily achievable whereas the conventional LOBS, i.e., group-I category produced from SDU process as mentioned above does not hold such restriction in sulphur and paraffin content limit and it is sold as per customer requirement of properties or international spec. for group-I LOBS for grades, 150N, 500N, 1300N, 150BS, etc., with varying viscosity, VI and pour point in different grade.

The operation in CIDW, being more on auto control and on more safety interlocks, is more or less simple and problems are also easy to tackle. The problems in CIDW are common like other hydrotreating, hydrocracking and isomerization processes as far as technology is concerned and hence not discussed to avoid repetition. Hence, problems are discussed in general with experiences instead of in the form of the specific case studies as given below:

6.14.1 Lower DWO Yield

Unlike other hydrotreating and isomerization process but like hydrocracking process, undesired cracking occurs thereby reducing yield of LOBS and generating some fuel distillates at the cost of DWO like in fuel hydrocracker sometimes more light distillates are generated at the cost of middle distillate as desired product. This problem in hydrocracker is already discussed in respective sections.

But to tackle this problem in CIDW, one has to produce narrow cut vacuum distillate to enhance DWO yield and reduce hydrocracked product as in hydrotreater reactor of CIDW, undesired hydrocracking is more than in fuel hydrotreater.

6.14.2 DWO VI Low, i.e., Off-Spec but Pour Point Very Low, i.e., Quality Give Away

Isomerization reactor severity to be reduced and third reactor of hydrogenation severity to be increased subsequently if required.

6.14.3 DWO VI Low and Pour Point is On-Spec

In this case, the first reactor of hydrotreating severity should be increased.

6.14.4 DWO VI High, i.e., Giveaway but Pour Point Marginal to Spec or Off-Spec

The first reactor of hydrotreating severity should be reduced and the second reactor of isomerization severity to be increased subsequently if DWO pours point is affected.

Common lesson: Problems occur mostly on account of feed quality changes as long as established operating conditions are not changed.

6.15 Solvent Deoiling Unit (SDU) for Wax Production from Vacuum Distillate

Here, only MIBK is used as solvent instead of using mix of MEK and toluene as deoiling is done with final chillers' outlet temperature of about +3 to +10°C as compared to −12 to −15°C in solvent dewaxing as explained earlier because if less chilling is done, MIBK with strong anti-solvent properties to wax and moderate solvent properties to oil is adequate as at higher temperature oil dissolving properties is enhanced where at the same time advantage of using single solvent is available to ease out the equipment design.

Conceptually, the problems in nature are the same as discussed in SDU earlier and the same mathematical correlation as shown in the solvent dewaxing section is applicable here also. Hence, troubleshooting separately is not discussed here except with one problem as observed during commissioning of a new wax deoiling unit as detailed below.

6.15.1 Start-Up of the Unit Could not be Established

- **Problem:** Like the solvent dewaxing process, during start-up circulation, solvent got chilled over a series of chillers and taken to recovery circuit bypassing filters. It was observed that filter bypass lines are small in size, i.e., not meant for flow during normal operation. Hence, during start-up chilling, less solvent flow was maintained. Also, the filters, though commissioned with cold wash spray, were not getting adequately cooled with such minimum flow of cold wash as usual. Hence, when feed cut is done, sudden load increase in chillers occurred due to not injection of feed but also increase in solvent flow through chillers thus not getting desired crystallization in chillers. Also, while filters were taken in line with feed-solvent mix, the wax crystals which were formed got destroyed in the filters as filters were not cooled enough with low cold wash flow beforehand.
- **Troubleshooting:** Start-up procedure was modified, i.e., start-up solvent chilling rate increased and this chilled solvent

was made to flow through filters under vacuum as if unit has been commissioned fully but without feed; when the whole unit normalized with only solvent flow, then feed cut in was done and good filter cake was observed as filters were in ready conditions with respect to temperature and vacuum. Thus operation got satisfactorily established.

- **Lesson:** Again it is an art that is learnt with experience. Hence, operating persons engaged in this unit should not be transferred elsewhere till his knowledge is adequately transferred.

6.16 Lube Hydrofinishing Unit (Lube HFU)

Lube HFU is a mild hydrotreatment unit for hydrofinishing DWO obtained from SDU to improve colour and oxidation stability of LOBS through partial desulphurization, denitrification and olefin hydrogenation respectively in presence of hydrogen in lube HFU. Here reaction occurs at system pressure of about 60 kg/cm^{2}g. Nitrogen contributes in colour more than sulphur. Also, LOBS obtained from this unit becomes transparent in colour.

Problems and troubleshooting experiences are the same as in other hydrodesulphurization as discussed earlier. However, one exceptional problem encountered in the unit is discussed below along with troubleshooting.

6.16.1 Product Ex Unit Found Hazy in Appearance

- **Problem:** There is no sulphur and nitrogen content limitation in product specifications but there are product colour specifications in ASTM scale which could be achieved by desulphurization and denitrification. As a normal protocol, there was one specification like "physical appearance" of the product which should be "transparent". But the product was hazy and thus was not sellable.
- **Troubleshooting:** LOBS in the post hydrotreatment recovery section passed through vacuum stripper to remove moisture, if any; conditions in vacuum stripper was absolutely normal; hence, there was no chance of product being hazy with

presence of moisture; moreover, when product sample on mild heating did not become transparent it indicated that it was not moisture which made product hazy.

The degree of desulphurization and denitrification was found from lab analysis and was compared with those found in the previous cycle of operation. It was found that the degree of desulphurization, denitrification has reduced though not that much to achieve product colour, but sulphur if not reduced up to a certain level, product should remain hazy. It should also be mentioned that unit catalyst life crossed 200% of guaranteed life but still operation was continued as there was no problem in getting product specifications. But this time, the reactor pressure drop was found to be excessively high and just above the allowable limit; with this situation, there would be every possibility of channelling inside the reactor. Moreover, make-up hydrogen purity also was very low at about 70% as against stipulated requirement of 75%.

To survive in the situation, an upstream CRU operating group was requested to improve H_2 purity. They could improve up to 73% after which product hazy problems get solved.

- **Lesson:** Here, with minor improvement in causes, the problem got solved, but as seen above operation could not last long with such high delta pressure across the reactor. Hence, catalyst replacement had to be done at the earliest.

6.17 Wax Hydrofinishing Unit

The purpose of this unit is same like Lube HFU as discussed above, i.e., colour improvement; but colour specification of Wax, particularly, for Food Grade Wax, is very stringent, i.e., product wax colour specification is in Lovibond scale and up to value of 0.1Y (maximum) in 18 Cell which indirectly indicates that product colour should be snow white. Hence, the degree of hydrotreating is more severe as compared to lube hydrofinishing. In a wax hydrofinishing unit (WHFU), reaction system pressure has to be maintained at 100–110 kg/cm^{2}g in presence of make-up hydrogen though not much hydrogen is consumed for full desulphurization and denitrification. But at high pressure and temperature of about 310°C in start of run

(SOR), aromatic (which contributes to colour) present if any gets hydrogenated, and also, asphaltene present, if any, gets trapped in catalyst pores to further contribute to colour improvement.

Common problems are discussed in general instead of case studies as follows.

6.17.1 Congealing Problem

It is a big threat for operation in this unit, and with effective heat tracing, the problem should be tackled.

6.17.2 Product Colour Problem

Sometimes it was seen that some colour stint remained in the product; it was due to presence of asphaltene which even in parts per billion (PPB) level deteriorates the colour; though temperature and pressure was adequate for hydrotreating, still temperature had to be further increased to facilitate increase of porosity in the catalyst just to trap the asphaltene and solve the problem. But this indicated that CCR of feed was high and should be reduced otherwise catalyst life would be reduced by premature increase of reactor temperature.

6.17.3 Product Oxidation Stability

With full desulphurization in these operating conditions, oxidation stability decreases as sulphur contributes to oxidation stability. Hence at product run down line of this unit, like in CIDU, anti-oxidant dosing is done to get back this property.

Many a time, neglecting operation monitoring of antioxidant dosing made the whole product tank off spec.

6.18 Hydrogen Generation Unit (HGU)

With presence of lot of hydrotreatment units, HGU unit installation for steam reforming of naphtha or natural gas (NG) to produce hydrogen and subsequently processing it in PSA unit to get 99% purity hydrogen becomes necessary as hydrogen ex CRU with purity of about 80% is not be adequate for all hydrotreatment processes.

HGU unit operation is mostly controlled with lots of safety interlocks and thus operation becomes simple. Problems beyond the identified ones are very less as far as reaction technology is concerned. However, two problems encountered are discussed as follows.

6.18.1 Pre-reformer Catalyst Activity Deteriorated Prematurely

While NG is used as feed, HGU can be operated with only reformer and without pre-reformer at upstream as NG contains about 92 to 94% methane and hence, pre-reformer is not required.

- **Problem:** The Z-90 plot of pre-reformer found to be at about 70% level just after a few months of operation. Z-90 plot indicates the location of the reactor where 90% of pre-reforming reaction, i.e., conversion of naphtha to methane gets achieved. With an expected catalyst cycle length of about minimum two years, Z-90 in 6 months of operation should not cross 25% of the reactor. So, this unwarranted incident caused a threat in HGU cycle length reduction.
- **Troubleshooting:** Feed, i.e., naphtha cut in pre-reformer should be done in between 490 and 500°C as above 500°C, naphtha gets cracked into fine cokes blocking catalyst pores. It was also found that on several occasions, the operating crew could not control pre-reformer bed temperature smoothly and when naphtha cut in done temperature went as high as up to 510–520°C; instead they should have decreased temperature again before feed cut in.
- **Lesson:** Though catalyst was replaced premature, this mal-operation did not recur in future because learning was well documented in start-up procedure.

6.18.2 Corrosion in Process Condensate Lines Post Reformer and Downstream Circuits

- **Problem:** The process condensate lines though made of stainless steel (SS) found to corrode during operation causing undesired unit shutdown with consequent forced shutdown of the hydrotreating units thereby causing production loss.

- **Troubleshooting:** There was a pH control chemical dosing system in the unit to improve the pH of process condensate up to neutral value of 7. The dosing system was working all right; still pH never came up to 7 and it could be at maximum achieved up to 6 and sometimes 5.5. It was found that oxalic acid generated during the reforming reaction contributed to this acidity. Based on this information, in the next cycle metallurgy was further improved in process condensate lines and the problem got solved.

6.19 Sulphur Recovery Unit

To meet the statutory stipulations on environment control, every refinery installed a sulphur recovery unit (SRU) to extract sulphur from refinery sour gases instead of flaring sour gas which results in emission of SO_2 to the atmosphere. To do so, refinery sour gases pass through an amine absorber unit (ARU), and sour waters pass through sour water stripper unit (SWS) to extract H_2S to process it in SRU. But in SWS, along with H_2S, NH_3 is also present. Earlier days, in SWS, single stripper was used, thus both H_2S and NH_3 together used to enter in SRU. As this NH_3 forms ammonium salt thereby clogging reboiler tubes in SRU, the temperature of main combustion chamber (MCC) of SRU had to be maintained at or above 1250°C at which NH_3 breaks into N_2 and H_2O to prevent ammonium salt deposition in reboiler tubes post MCC. But now a day, in SWS, two stage stripping is done with first one operating at high pressure of about 7 kg/cm^{2}g from where only H_2S is released, and the second stage is operated at low pressure of about 0.5 kg/cm^2 to release NH_3 and may be with traces of H_2S which is sent to incinerator to burn for emission to atmosphere. In such a situation, MCC temperature can be maintained at about 700°C, i.e., only to ensure partial clause reaction to happen for conversion of H_2S to elemental sulphur. But still, there are lots of problem in operating SRU as discussed below:

6.19.1 Ingress of Hydrocarbon along with H_2S

- Problem: Ingress of hydrocarbon along with H_2S not only creates generation of poison like carbonyl as poison of Clause catalyst, but also may generate coke to reduce catalyst activity.
- **Solution:** This should be prevented by taking refinery sour water to a feed tank in SWS instead of directly feeding into it to allow settling of this oil and then taking oil free water into the stripper. However, best control should be to prevent oil escape into sour water in the upstream units.

  Notes: Other problems on impurities like chloride, phosgenes also deactivate the catalyst and care should be taken as per manual procedure.

6.19.2 Congealing of Condensers' Outlet Lines

- **Problem:** Condenser outlet lines found congealed when the unit started after shutdown.
- **Solution:** Condensers outlet lines are steam jacketed; hence, leak, if any, inside the jacket cannot be seen. It was found that liquid sulphur line had leaked on corrosion and water from the jacket entered into the sulphur line thereby cooling the line during total shutdown instead of keeping it heated. Adequate protection in operating conditions and condensers outlet lines' metallurgy should be taken to prevent it.

6.19.3 Sulphur Quality Issue

- **Problem:** Sulphur quality sometimes found gets affected like colour and physical structure.
- **Solution:** Generally this does not happen; as an exception it may happen only due to undesired polymerization occurring in run sulphur pumping chambers. To prevent this polymerization, suitable chemical dosing should be done and also sulphur locks should be kept properly tight.